Material Agency

Carl Knappett · Lambros Malafouris
Editors

Material Agency

Towards a Non-Anthropocentric Approach

Springer

Editors
Carl Knappett
University of Toronto, Toronto,
Ontario, Canada

Lambros Malafouris
University of Cambridge
Cambridge, UK

ISBN: 978-0-387-74710-1 e-ISBN: 978-0-387-74711-8
DOI: 10.1007/978-0-387-74711-8

Library of Congress Control Number: 2008930774

Printed on acid-free paper

springer.com

Contents

Contributors

Andy Clark
Department of Philosophy, Edinburgh University, Edinburgh, UK

Paul Cloke
Department of Geography, University of Exeter, Exeter, UK

Andrew Cochrane
School of History and Archaeology, University of Cardiff, Cardiff, Wales, UK

Richard Harper
Microsoft Research Cambridge, Cambridge, UK

Tim Ingold
Department of Anthropology, School of Social Science, University of Aberdeen, Aberdeen, UK

Owain Jones
Countryside and Community Research Institute, Dunholme Villa, The Park, Cheltenham, UK

Carl Knappett
Department of Art, University of Toronto, Toronto, Ontario, Canada

John Law
Centre for Science Studies and Department of Sociology, Lancaster University, Lancaster, UK

Lambros Malafouris
McDonald Institute for Archaeological Research, Cambridge University, Cambridge, UK

Annemarie Mol
Centre for Science Studies and Department of Sociology, Lancaster University, Lancaster LA1, 4YN, UK

Micheal Molloy
Microsoft Research Cambridge, Cambridge, UK

John Sutton
Macquarie Centre for Cognitive Science, Macquarie University, Sydney, NSW 2109, Australia

Alex Taylor
Microsoft Research Cambridge, Cambridge, UK

Sander van der Leeuw
School of Human Evolution and Social Change, Arizona State University, Tempe, Arizona, US

Christopher Watts
Department of Anthropology, University of Toronto, Toronto, Ontario, M5S 2S2, Canada

Tom Yarrow
Department of Social Anthropology, School of Social Sciences, University of Manchester, Manchester, UK

Material and Nonhuman Agency: An Introduction

C. Knappett and L. Malafouris

Material and nonhuman agency – surely this is a mistake? Is not agency a solely human property? How then can we devote a whole volume to a topic with such obviously shaky foundations? Certainly, the odds seem to be stacked against us when we think of agency as not only the capacity to act, but also the capacity to reflect on this capacity. A subject may feel his or her arm moving and recognise 'ownership' of that movement, but this is not necessarily the same as being able to reflectively understand that he or she is the cause or 'agent' of that movement (Gallagher 2007, p. 2). When agency is linked strictly to consciousness and intentionality, we have very little scope for extending its reach beyond the human.

Even those nonhuman entities that seem to threaten this neat equation most – let us say software agents or robotic agent-artefacts – do not get close to fulfilling these criteria of agency, even though they are closely modelled after the human (Suchman 2007; Sørensen and Ziemke 2007). In this view, then, 'material agency' is a secondary property, a mirage even, with agency (as consciousness and intentionality) still very much in human hands. Nonetheless, there is some sense, even an anxiety in folk psychology, that the autonomy and interactivity of such artefacts is a step towards agency. But achieving agency is from this perspective a question, essentially, of becoming human.

This human-centred view of agents and artefacts is not limited to those artefacts we design to be like agents. It extends to a much wider and more prosaic world of artefacts and matter, an environment of things that is conceived on our own terms, under our control and designed to serve. We do not give a second thought, on the whole, to chairs, mugs, steps, litterbins, wooden, ceramic, concrete or plastic: these objects are overlooked because we engage with them habitually and haptically every day. They would not serve our ends very well, if we could not overlook them. Designed to be secondary, they have to be secondary, forming the backdrop to our lives, of which we are of course the stars, the decision-makers, the agents. It is common sense that agency should be conceived anthropocentrically – how can it be otherwise? We are centre-stage in our lives, not these artefacts, however mundane, or indeed intelligent.

Although here glossed rather simplistically, this anthropocentric worldview means that the material or environmental counterpoints to human agency have generally been given short shrift in scholarly discussion. Indeed, while agency is a much-debated theme across the social sciences, the terms of the debate have remained rather narrow, focussing overwhelmingly on the relationship between agency and structure (Emirbayer and Mische 1998, pp. 962–963). Arguments go back and forth over the degree to which agents, which may be, though certainly need not be, human individuals, are free to act in the world. This degree of freedom is socio-culturally mediated (Ahern 2001, p. 109), and thus varies considerably in different settings and societies. It is a fundamental question in how society is constituted, from the top-down or the bottom-up: does power lie with individuals, or with social institutions? This debate, often cast in terms of structure vs. agency, has seen attempts to bridge the divide, in the form of practice theory and structurationist ideas (Archer 1988; Bourdieu 1977, 1990; Giddens 1984). Such ideas are now familiar territory not only across much social science, but also in archaeology, with the work of Bourdieu and Giddens frequently cited (Barrett 2001; Dornan 2002; Gardner 2004; Johnson 2000; Smith 2001). Agency has become something of a buzzword in archaeological theory (Dobres and Robb 2000, p. 3), particularly in post-processual thinking, where it is a byword for individual rather than systemic power in the past.[1]

Yet while the concept of agency is much contested, it is done so within the theoretical margins of a narrow anthropocentric perspective. This anthropocentric view of agency is based upon a general agreement about a single undisputable fact: that agency, in the real sense of the word, is a property of the human individual – "the only true agents in history are human individuals" (Giddens and Pierson 1998, p. 89). Whether this individual is conceived through a Cartesian or an existential lens makes no important difference. What is important is that when we speak about agents *proper*, we are referring to human individuals, and preferably of the modern Western-type. In short, agency is an attribute of the human substance.

However, an emergent suspicion of the humanistic determinations of agency can actually be traced quite far back, to the likes of Mauss (1954) and Heidegger (1977). Mauss's seminal study, *The Gift*, illustrated the fluidity of the boundaries between persons and things and the capacity of the latter to embody and objectify, as well as produce, social consequences. This is a point also reflected in Munn's observations on the Kula exchange system: "Men appear as the agents defining shell value, but without shells men cannot define their value. In this respect, shells and men are reciprocally agents of each other's value definition", (Munn 1983, p. 284). Indeed, as modern anthropology has plainly illustrated "commodities, like persons, have social lives" (Appadurai 1986, p. 3). The *enframing* that, according to Heidegger, characterises the attitude of the

[1] "[A]gency has become the buzzword of contemporary archaeological theory ... a lingua franca – an ambiguous platitude meaning everything and nothing ... there is little consensus about what "agency" actually means ... nor has there been sustained consideration of basic methodological and epistemological issues as to make it applicable and appropriate to the premodern past" (Dobres and Robb 2000, p. 3).

Western individual towards the world as a "standing reserve" – a passive recourse to be controlled and manipulated for human ends – has no place and meaning in a number of ethnographic contexts with a very different understanding of what is it like to be a person, and what it means to engage the world (Heidegger 1977). For example, what could agency mean to a partible, composite and relationally constituted Melanesian person (Strathern 1988)? Clearly the idea of the isolated agent that acts upon the world, imposing shape and meaning upon inert matter, can hardly be accommodated in a Melanesian context where the categories of persons and things are inseparably distributed over biographical time and space.

Despite those examples, however, it has been primarily in the last two decades that the idea of decentralised agency has gained momentum across the social sciences. It is Actor-Network Theory (ANT) that has been particularly influential in this regard (Callon 1986; Law 1992, 1999, 2002; Latour 1994, 1999a, b, 2000, 2005). ANT can be defined as a semiotics of materiality that is symmetrical with respect to human and nonhuman agents (Law 1999, p. 4). Conceptualizing agency as variously distributed and possessed in relational networks of persons and things, ANT proposes that all entities participating in those networks should be treated analytically as of equal importance (Ashmore et al. 1994; Fuller 1994; Lee and Brown 1994). In other words, for ANT what we call actors or agents are essentially the products or effects of networks. That means that no primacy of the human actor – individual or collective – over the nonhuman actor can be accepted on a priori grounds. This may sound like yet another attempt to reconcile the two traditionally opposed poles of social theory, i.e. agency and structure, but in reality is something quite different. In drawing material things into the sociological fold the aim of ANT was not to overcome this contradiction but simply to ignore it, and develop what Latour calls a "bypassing strategy" (1999, pp. 16–17).

For example, to answer the question whether people or guns kill, we need to move beyond what is acceptable in either the materialist or the sociological account of activity (Latour 1999, p. 180). Both accounts start with essences, and essences result in antinomies, and antinomies are the reason that modernist theories fail to capture practice.

> What does the gun add to the shooting? In the materialist account, *everything*: an innocent citizen becomes a criminal by virtue of the gun in her hand. The gun enables, of course, but also instructs, directs, even pulls the trigger ... Each artifact has its script, its potential to take hold of passerby and force them to play a role in its story. By contrast, the sociological version ... renders the gun a neutral carrier of will that adds nothing to the action, playing the role of a passive conductor, through which good and evil are equally able to flow (Latour 1999, p. 177).

What both accounts – materialist and sociological – fail to recognise is that agency "resides in the blind spot in which society and matter exchange properties" (Latour 1999, p. 190). Neither the isolated gun nor the isolated individual can bear the responsibility for the act of killing. The responsibility lies, on the one hand, in the way those two agents come together to construct a new hybrid agent – the gunman – and on the other, in the socio-technical network that supports and makes possible such a meeting. Action involves a coalescence of human and nonhuman elements and as such the responsibility for action must be shared among them (Latour 1999, pp. 180–182). No distinctions between human and nonhuman entities can be sustained in terms of agency.

More will be said on ANT in subsequent chapters, such as those by Knappett and Watts. For similar perspectives, we might also consider Pickering's (1995) work on "the dance of agency" between humans and artefacts, Kaufmann's "la danse avec les choses" (1997), or Suchman's "sociomaterial agency" (2007). Cognitive science and philosophy have also advanced our understanding of the agency of artefacts in novel ways (see Hutchins 1995; Kirsh 1995; Clark 1997; Norman 1988). We might also very well turn to other domains such as human geography (Jones and Cloke 2002), political theory (Bennett 2004), economics (Lane and Maxfield 1997) and anthropology (Gell 1998; Layton 2003; Hoskins 2006). Many of these perspectives are more concerned with understanding agency as a situated process, rather than debating what or who is or is not an agent. This has, in part, been aided by a recognition that agency need not be coterminous with intentionality, which releases nonhumans into the process of agency. We hope to capture in this volume some of this move away from anthropocentric approaches that is happening across the social sciences. We also intend to show the worth of these new perspectives for archaeology, which has been slow on the uptake; despite the fact that the likes of Bourdieu and Giddens show little concern for material culture; much archaeological theory has remained faithfully wed to practice theory and structuration for 20 years. By using the term "material agency" we do not want to go to the other extreme and say that agency is material rather than human; it is more of a wake-up call, for social scientists and archaeologists, to encourage them to consider agency non-anthropocentrically, as a situated process in which material culture is entangled. Archaeology has the potential here to lead the way, as it deals with material culture "far more seriously and innovatively than do most social scientists" (Dobres and Robb 2000, p. 14).

Given this obvious focus on material culture in archaeology, and the acknowledgement that "material culture is clearly central to creating agents and expressing agency" (Dobres and Robb 2000, p. 14), it is surprising that the relationship between material culture and agency does not feature more prominently in current archaeological theory.[2] Strange as it might seem for a discipline that is in broad agreement on the "active" nature of material culture, archaeology remains, in our opinion, attached to an *anthropocentric* view of the world and by extension also of agency. The move towards a more active view of

---

[2] In a recent review, Dobres and Robb note that agency remains both "woefully under-theorised" and subject to remarkably "sparse methodological developments" (2005, p. 159). While Dobres and Robb stress in particular the need for robust methodologies in the study of agency (see also Joyce and Lopiparo 2005), we focus here more on the woeful under-theorisation.

material culture goes back 20 years or more; but this has been more a case of acknowledging the active rather than passive use of material culture by humans, rather than ascribing much dynamism to the artefacts themselves. What the active nature of material culture in its common usage seems to imply is essentially that *human* individuals, far from passively adapting to external systemic forces, are *actively using* material culture as an expressive symbolic medium for their social strategies and negotiations (Hodder 1982, 1986). In other words, the essence of the argument is that material culture may not simply reflect but also actively construct or challenge social reality, *on the necessary condition,* however, of human agency and intentionality. The above sounds too obvious to be wrong, and indeed this is precisely how material culture operates in many cases. However, this is only a part of the picture and, once you embrace it, leaves you with few chances to discover what the active nature of material culture really means.

There are some recent exceptions that do take the activeness of artefacts themselves more seriously, but they are surprisingly few (Chapman 2000; Chapman and Gaydarska 2007; Gamble 2007; Olsen 2003; Shanks 1998; Webmoor and Witmore 2005). Perhaps we should look beyond archaeology, particularly given this volume's interdisciplinary breadth; if we turn to the emerging field of "material culture studies", does this fare much better? A good place to assess this is the *Journal of Material Culture*, founded in 1996. A search reveals the word "agency" mentioned in the texts of 111 papers over the last 10 years. While this might at first glance appear to represent a strong commitment to issues of agency across a range of approaches in material culture studies, there is a remarkable tendency in the vast majority of these cases to engage with agency in a very particular way, influenced predominantly by Alfred Gell's 1998 volume "Art and Agency". Thus even here the concept of agency is not subject to very much depth or breadth of investigation.

These modest advances notwithstanding, we believe that approaches to material agency would benefit greatly from a much broader interdisciplinary involvement. This is one of the main justifications for this volume.

We begin with a contribution from Andy Clark, one of the foremost philosophers of mind to have contributed to discussions of agency. In "Where Brain, Body and World Collide" (Chapter 2), which was originally published in Daedalus in 1998, Clark sets out a possible conceptual background and philosophical conception of self and the mind with which many of the independent perspectives and case studies presented in this volume can be anchored. Drawing on recent theoretical and experimental work ranging from monkey finger control and mirror neurons to interactive vision and robotics, Clark puts together a powerful argument for the integration of perception, cognition and action in the study of mind and agency. For him the agency of the material world is simply the natural consequence of a "deeply interanimated unity" between perception and action.

This view of the situated brain and the extended mind is then explored by Malafouris using the potter's wheel as his example (Chapter 3). Developing his argument for Material Agency, Malafouris argues that the only available starting point and obligatory point of passage for studying the emergence and determination of agency is that of material engagement. As with many other dimensions of the human mind, agency and intentionality should be understood as distributed, emergent and interactive phenomena rather than as subjective experiences. The clay on the potter's wheel should not be construed as the external passive object of the potter's intentional states, but as a functionally co-substantial component of the intentional character of the potting experience.

Our trio of cognitive science approaches to agency is completed by John Sutton. His chapter "Material agency, skills, and history" (Chapter 4) explores the issues of agency, interactivity, skill and distributed cognition from a broader historic and diachronic perspective. How can we identify the significant dimensions of cross-cultural and historical variation that will enable the construction of better typologies of distributed cognitive systems? What might be the role of cognitive archaeology to this end? How can archaeology and anthropology broaden the empirical and theoretical horizons of the cognitive sciences? Taking issue with a recent paper by one of us (Malafouris 2004), and focusing on the agency of things in the context of memory, Sutton proposes a number of fruitful directions for interdisciplinary research and the cross-fertilisation of ideas.

The next three chapters present diverse interdisciplinary perspectives on material agency, ranging from sociology and human geography, through to economics and human–computer interaction. The first two chapters, those of Law and Mol, and Jones and Cloke, provide us with an important correction to the potential pitfall in a non-anthropocentric approach of focussing only on artefacts and technologies. They show us that "material" in this case means "nonhuman", with their case studies tackling sheep and trees, respectively. Law and Mol (Chapter 5) take a Cumbrian sheep as their prospective agent, specifically a Cumbrian sheep in the midst of the foot and mouth crisis of 2001. Situating their approach within the "material semiotics" afforded by ANT, they stress that entities give each other being: they both act and are "enacted" (hence their chapter title "The Actor-Enacted"). They allow for four versions of a sheep – as a veterinary, epidemiological, economic and farming entity. These entail multiple practices that enact the sheep, making for a "sheep multiple". However, an enacted sheep is not a passive sheep; but it is difficult in the English language to circumvent this active/passive dichotomy. Though difficult to imagine a simultaneously active–passive agent, this is precisely what is required if we are to understand sheep as actors in this scenario. Furthermore, the practices that enact sheep-actors form complex webs that merge, interfere and pull apart, dynamically and indeterminately; we would be better advised, say Law and Mol, to ask *what* is happening, not *who* has done it.

The chapter by Jones and Cloke (Chapter 6) also falls under the broad ANT umbrella, with its presentation of trees as "actants", and agency as a "hybridised" phenomenon. They note, however, that ANT, in its discussions of hybridity, has been "biased towards technological rather than organic nonhuman entities". Not only this, but when the organic is included, it ventures only as "far" as animals. Thus Jones and Cloke set out to remedy this imbalance in material agency by looking at other kinds of organism, in this case trees. They take three "tree places" in and around Bristol (South-west England) and assess the agency of trees in the making of these places over time. In so doing they distinguish between four kinds of agency, as routine, transformative, purposive and non-reflexive action.

In the approach taken by Harper, Taylor and Molloy (Chapter 7), we see the kinds of contemporary and future human–computer interaction that might easily tempt us into assigning agency to artefacts that appear somehow "intelligent". Yet the authors are quick to state that the objects in question, though dynamic and interactive, do not have intelligence. This they assign to humans alone. They show how surfaces and containers in the home have particular affordances that can be augmented with digital technologies; but surfaces remain just surfaces, and do not become artefacts of intelligence through digital augmentation.

In a third section of the volume, we have four chapters dealing with archaeological contexts, albeit from interdisciplinary perspectives. Yarrow's contribution (Chapter 8) is actually more an ethnography of recording processes during archaeological excavation. He looks in particular at context sheets, typically thought of as passive records of archaeological features but which, Yarrow argues, do have an enactive role within wider networks which include both human and nonhuman actants. There are interesting resonances with some of Harper's earlier work on the role of paper documents in offices.

Chapter 9 by Knappett picks up on the theme of networks of human and nonhuman actants, seeking to develop more systematic means for addressing the network properties of agency, paradoxically neglected in Actor-*Network* Theory. Using examples from the Aegean Bronze Age, he examines the interactions between different categories of "material culture" – artefacts, images and texts. Early Aegean scripts are to some extent imagistic, and indeed they include images of artefacts; and some texts take artefactual form, inscribed on various kinds of support (e.g. tablets, sealings and pots). By looking at these various connections, we can gain glimpses into how artefacts, images and texts had a networked presence in the world of the Aegean Bronze Age; and that this presence formed a kind of material agency within which humans were deeply entwined.

Following this comes a contribution on the intertwined agency of water and stone in Irish Neolithic passage tombs (Chapter 10). Andrew Cochrane argues that the imagery carved into the stone façades of these tombs was "stimulated" by solutions, i.e. when wetted by rainwater and perhaps other kinds of liquid. That these images might come to life when wet, stimulating new perceptions and experiences among their viewers, is very suggestive; we might see some parallels with the paper of Harper et al., in that the affordances of surfaces and substances might be exploited and augmented by designers and/or users. Once again, it may not be easy to locate agency in this scenario; and we might follow Law and Mol's advice in asking *what* is happening in Cochrane's stone-and-solution scenario rather than *who* has done it.

The final Chapter 11 in this archaeological section comes from Chris Watts, who grapples with the complexities of Peircean semiotics, or 'semeiotic', in developing what one might call a situated semiotic approach to material agency. His use of Peirce's notion of "synechism" conveys very persuasively the way in which "people and things are conjoined through the principle of semiotic mediation". Watts grounds his sophisticated theoretical arguments in a case study drawn from the Late Woodland period of southwestern Ontario; comparing Iroquoian with Western Basin ceramic assemblages, he is able to show how these traditions bring into being two quite different agentic networks.

The volume concludes with two sets of concluding remarks, by Tim Ingold (Chapter 12) and Sander van der Leeuw (Chapter 13), respectively. The first of these is an ingenious critique of the ANT approach as applied to questions of material agency. Ingold creates a metaphor whereby "ANT" is one kind of creature in the forest, and "SPIDER" is another, and they strike up a philosophical dialogue. We know that ANT stands for "Actor Network Theory", but SPIDER is a new acronym devised by Ingold that stands for "Skilled Practice Involves Developmentally Embodied Responsiveness". Whereas ANT sees agency as coming about through the networked interlinkage of diverse objects, as for example in an ant colony, SPIDER rather sees agency as emergent from the skilled action-perception of an organism that inhabits a particular milieu, as for example a spider-with-web. The skilled action-perception of an organism as it moves through an environment creates a kind of mesh, but this does not mean that agency is distributed evenly between the organism and its mesh (which could be a fish with water, a butterfly with air, a spider with web or a potter with clay). Ingold thereby provides us with a thought-provoking challenge to ANT, particularly its tendency to see agency as a distributed network phenomenon, without sufficient attention to either the role of bodily perception and movement in the creation of the network, or the different qualities of the entities that are implicated in this process. Ingold's wider aim is to get us thinking in terms of "meshworks" rather than "networks" (see Ingold 2007).

Sander van der Leeuw's concluding comments (Chapter 13) have a different flavour. He draws on his experience of long-term, deep-rooted interdisciplinary projects in assessing the overall impact of the papers in this volume, both for archaeology and the social sciences more broadly. At the same time, he considers their potential for the development of a new approach to invention and innovation, topics of particular interest to van der Leeuw. He differentiates between a priori and a posteriori perspectives on innovation, the former engendering a proactive approach and the latter a reactive one. Unfortunately, the tendency is towards the latter in much of the scientific writing on innovation, which means that the inventive creativity leading to innovation is poorly understood. Van der Leeuw thus takes this volume's emphasis on relationality (established through network thinking) to rethink the relationship between invention and innovation, with reference to traditional pottery making in Mexico and the Philippines. He sees the potential for inventive action as a feature of the dynamic "network", composed of people, things, objects and contexts, within which a potter is situated. Invention is a local process, only involving a relatively restricted network, whereas innovation is more widespread, cascading through a broader network.

Finally, we should emphasise that the aim of this volume is not to establish "material" or "nonhuman" agency as some kind of sustainable alternative to the idea of human agency. Rather, our primary intention is to provoke debate, with the term "material agency" intended as a challenge to the anthropocentrism inherent in existing approaches to agency. Some of the contributions to this volume argue that it is correct to centre our understanding of agency on the human; we are not averse to this when it is clearly and explicitly argued, as it is here, rather than implicitly assumed. We have not been looking for a consensus, but have actively sought divergent views. Such divergence is all the more likely when a project cuts across disciplinary lines, as here in the apparently unlikely combination of computer science, cognitive science, philosophy, human geography, and sociology, as well as anthropology and archaeology. Nevertheless, we do find that there are many converging lines of enquiry too, which here we have only begun to broach, and which, we believe, merit much fuller exploration in the future.

References

Ahern, L., 2001, Language and Agency, *Annual Review of Anthropology* 20: 109–137.

Appadurai, A., 1986. Introduction: Commodities and the Politics of Value, In *The Social Life of Things*, edited by A. Appadurai, pp. 3–63. Cambridge: Cambridge University Press.

Archer, M. S., 1988, *Culture and Agency: The Place of Culture in Social Theory*. Cambridge University Press, Cambridge.

Ashmore, M., Woofitt, R., and Harding, S., 1994, Humans and Others, Agents and Things, *American Behavioral Scientist* 37(6): 733–740.

Barrett, J., 2001, Agency, the Duality of Structure, and the Problem of the Archaeological Record. In *Archaeological Theory Today*, edited by I. Hodder, pp. 141–164. Polity Press, Cambridge.

Bennett, J., 2004, The Force of Things: Steps Towards an Ecology of Matter, *Political Theory* 32(3): 347–372.

Bourdieu, P., 1977, *Outline of a Theory of Practice*. Cambridge University Press, Cambridge.

Bourdieu, P., 1990, *The Logic of Practice*. Polity Press, Cambridge.

Callon, M., 1986, Some Elements for A Sociology of Translation: Domestication of the Scallops and the Fishermen of St-Brieuc Bay. In *Power, Action and Belief: A New Sociology of Knowledge?*, edited by J. Law, pp. 196–223. Sociological Review Monograph, Routledge and Kegan Paul, London.

Chapman, J., 2000, *Fragmentation in Archaeology: People, Places and Broken Objects in the Prehistory of South-eastern Europe*. Routledge, London.

Chapman, J., and Gaydarska, B., 2007, *Parts and Wholes: Fragmentation in Prehistoric Context*. Oxbow Books, Oxford.

Clark, A., 1997, *Being There: Putting Brain, Body and World Together Again*. MIT Press, Cambridge, MA.

Dobres, M.-A., and Robb, J., 2000, Agency in Archaeology: Paradigm or Platitude? In *Agency in Archaeology*, edited by M.-A. Dobres and J. Robb, pp. 3–17. Routledge, London.

Dobres, M.-A., and Robb, J., 2005, "Doing" Agency: Introductory Remarks on Methodology, *Journal of Archaeological Method and Theory* 12(3): 159–166.
Dornan, J. L., 2002, Agency and Archaeology: Past, Present, and Future Directions, *Journal of Archaeological Method and Theory* 9(4): 303–329.
Emirbayer, M., and Mische, A., 1998, What is Agency? *American Journal of Sociology* 103(4): 962–1023.
Fuller, S., 1994, Making Agency Count: A Brief Foray into the Foundations of Social Theory, *American Behavioral Scientist* 37(6): 741–753.
Gallagher, S., 2007, The Natural Philosophy of Agency, *Philosophy Compass* 2 (j.1747-9991. 2006.00067.x).
Gamble, C., 2007, *Origins and Revolutions: Human Identity in Earliest Prehistory*. Cambridge University Press, Cambridge.
Gardner, A., editor, 2004, *Agency Uncovered: Archaeological Perspectives on Social Agency, Power and Being Human*. UCL Press, London.
Gell, A., 1998, *Art and Agency: Towards a New Anthropological Theory*. Clarendon Press, Oxford.
Giddens, A., 1984, *The Constitution of Society: Outline of the Theory of Structuration*. Polity Press, Cambridge.
Giddens, A., and Pierson, C., 1998, *Conversations with Anthony Giddens: Making Sense of Modernity*. Stanford University Press, Stanford, CA.
Heidegger, M., 1977. The Question concerning Technology and Other Essays. New York (NY): Harper and Row.
Hodder, I., 1982, *Symbols in Action. Ethnoarchaeological Studies of Material Culture*. Cambridge University Press, Cambridge.
Hodder, I., 1986, *Reading the Past*. Cambridge University Press, Cambridge.
Hoskins, J., 2006, Agency, Biography and Objects. In *Handbook of Material Culture*, edited by C. Tilley, W. Keane, S. Küchler, M. Rowlands and P. Spyer, pp. 74–84. Sage, London.
Hutchins, E., 1995, *Cognition in the Wild*. MIT Press, Cambridge, MA.
Ingold, T., 2007, *Lines: A Brief History*. Routledge, London.
Johnson, M., 2000, Conceptions of Agency in Archaeological Interpretation. In *Interpretive Archaeology: A Reader*, edited by J. Thomas, pp. 221–227. Leicester University Press, London.
Jones, O., and Cloke, P., 2002, *Tree Cultures: The Place of Trees and Trees in Their Place*. Berg, Oxford.
Joyce, R. A., and Lopiparo, J., 2005, PostScript: Doing Agency in Archaeology, *Journal of Archaeological Method and Theory* 12(4): 365–374.
Kaufmann, J.-C., 1997, *Le coeur à l'ouvrage: théorie de l'action ménagère*. Nathan, Paris.
Kirsh, D., 1995, The Intelligent Use of Space, *Artificial Intelligence* 73: 31–68.
Lane, D. A., and Maxfield, R., 1997, Foresight, Complexity and Strategy. In *The Economy as an Evolving Complex System 2*, edited by W. B. Arthur, S. N. Durlauf and D. A. Lane, pp. 169–198. Addison-Wesley, Reading, MA.
Latour, B., 1994, Pragmatogonies: A Mythical Account of How Humans and Nonhumans Swap Properties, *American Behavioral Scientist* 37(6): 791–808.
Latour, B., 1999a, *Pandora's Hope: Essays on the Reality of Science Studies*. Harvard University Press, Cambridge, MA.
Latour, B., 1999b, On Recalling ANT. In *Actor Network Theory and After*, edited by J. Law and J. Hassard, pp. 15–25. Blackwell, Oxford.
Latour, B., 2000, The Berlin eyK or How to do Words with Things. In *Matter, Materiality and Modern Culture*, edited by P. M. Graves-Brown, pp. 10–21. Routledge, London.
Latour, B., 2005, *Reassembling the Social: An Introduction to Actor-Network-Theory*. Oxford University Press, Oxford.
Law, J., editor, 1992, *A Sociology of Monsters: Essays on Power, Technology and Domination*. Routledge Sociological Review Monograph, London.

Law, J., 1999, After ANT: Complexity, Naming and Topology. In *Actor Network Theory and After*, edited by J. Law and J. Hassard, pp. 1–14. Blackwell, Oxford.
Law, J., 2002, *Aircraft Stories: Decentring the Object in Technoscience*. Duke University Press, Durham.
Layton, R., 2003, Art and Agency: A Reassessment, *Journal of the Royal Anthropological Institute* (n.s.) 9: 447–464.
Lee, N., and Brown, S., 1994, Otherness and the Actor Network, *American Behavioral Scientist* 37(6): 772–790.
Malafouris, L., 2004. The Cognitive Basis of Material Engagement: Where Brain, Body and Culture Conflate. In *Rethinking Materiality: The Engagement of Mind with the Material World*, edited by E. DeMarrais, C. Gosden and C. Renfrew, PP. 53-62. McDonald Institute Monographs, Cambridge.
Mauss, M., 1954. The Gift. London: Cohen and West.
Munn, N., 1973. Walpiri Iconography. Ithaca and New York: Cornell University Press.
Norman, D. A., 1988, *The Psychology of Everyday Things*. Basic Books, New York.
Olsen, B., 2003, Material Culture after Text: Re-Membering Things, *Norwegian Archaeological Review* 36(2): 87–104.
Pickering, A., 1995, *The Mangle of Practice: Time, Agency and Science*. Chicago University Press, Chicago.
Shanks, M., 1998, The Life of an Artifact, *Fennoscandia Archeologica* 15:15–42.
Smith, A. T., 2001, The Limitations of Doxa: Agency and Subjectivity from an Archaeological Point of View, *Journal of Social Archaeology* 1(2): 155–171.
Sørensen, M. H., and Ziemke, T., 2007, Agents Without Agency? *Cognitive Semiotics* (special issue), 102–124.
Strathern, M., 1988, *The Gender of the Gift*. University of California Press, Berkeley, CA.
Suchman, L., 2007, *Human-Machine Reconfigurations: Plans and Situated Actions, 2nd edition*. Cambridge University Press, Cambridge.
Webmoor, T., and Witmore, C., 2005, *Symmetrical Archaeology*, Metamedia, Stanford University @ traumwerk.stanford.edu:3455/Symmetry/home, accessed *26-11-06*.

Chapter 1
Where Brain, Body and World Collide

Andy Clark

Introduction The brain fascinates because it is the biological organ of mindfulness itself. It is the inner engine that drives intelligent behaviour. Such a depiction provides a worthy antidote to the once-popular vision of the mind as somehow lying outside the natural order. But it is a vision with a price. For it has concentrated much theoretical attention on an uncomfortably restricted space; the space of the inner neural machine, divorced from the wider world which then enters the story only via the hygienic gateways of perception and action. Recent work in neuroscience, robotics and psychology casts doubt on the effectiveness of such a shrunken perspective. Instead, it stresses the unexpected intimacy of brain, body and world and invites us to attend to the structure and dynamics of extended adaptive systems – ones involving a much wider variety of factors and forces. Whilst it needs to be handled with some caution, I believe there is much to be learnt from this broader vision. The mind itself, if such a vision is correct, is best understood as the activity of an essentially *situated* brain: a brain at home in its proper bodily, cultural and environmental niche.

Software

Humans, dogs, ferrets – these are, we would like to say, mindful things. Rocks, rivers and volcanoes are not. And no doubt there are plenty of cases in between (insects, bacteria, etc....). In the natural order, clear cases of mindfulness always involve creatures with brains. Hence, in part, the fascination of the brain, understanding the brain looks crucial to the project of understanding the mind. But how should such an understanding proceed?

An early sentiment – circa 1970 and no longer much in vogue – was that understanding the mind depended rather little on understanding the brain. The brain, it was agreed, was in some sense the physical medium of cognition. But

A. Clark
Department of Philosophy, Edinburgh University, Edinburgh, UK
e-mail: Andy.Clark@ed.ac.uk

C. Knappett, L. Malafouris (eds.), *Material Agency*,
DOI: 10.1007/978-0-387-74711-8_1, © Springer Science+Business Media, LLC 2008

everything that mattered about the brain *qua* mind-producing engine turned not on the physical details but on the computational and information-processing strategies that the neural stuff (merely) implemented. There was something importantly right about this view, but something desperately wrong as well.

What was right was the observation that understanding the physical workings was not going to be enough; that we would need also to understand how the system was organised at a higher level in order to grasp the roots of mindfulness in the firing of neurons. This point is forcefully made by the cognitive scientist Brian Cantwell Smith who draws a parallel with the project of understanding ordinary computer systems. With respect to, for example,, a standard PC running a tax-calculation program, we could quite easily answer all the "physiological" questions (using source code and wiring diagrams) yet still lack any real understanding of what the program does or even how it works.[1] To really understand how mental activity yields mental states, many theorists believe, we must likewise understand something of the computational/information-processing organisation of the brain. Physiological studies may contribute to this understanding. But even a full physiological story would not, in and of itself, reveal how brains work *qua* mind-producing engines.[2]

The danger, of course, was that this observation could be used as an excuse to downplay or marginalise the importance of looking at the biological brain *at all.* And so, in the early days of cognitive science, it was common to hear real neuroscience dismissed as having previous little to offer to the general project of understanding human intelligence. Such a dismissal, however, could not long be sustained (for a powerful critique, see Reeke & Edelman (1988)). For although it is (probably) true to say that a computational understanding is in principle independent[3] of the details of any specific implementation in hardware (or wetware), the project of *discovering* the relevant computational description (especially for biological systems) is surely not.

One key factor is evolution. Biological brains are the product of biological evolution and as such often fail to function in the ways we (as human designers) might expect.[4] This is because evolution is both constrained and liberated in ways we are not. It is constrained to build its solutions incrementally via a series of simpler but successful ancestral forms. The human lung, to give one example, is built via a process of tinkering (Jacob (1977)) with the swim bladder of the fish. The engineer might design a better lung from scratch. The tinkerer, by contrast, must take an existing device and subtly adapt it to a new role. From

[1] See B.C. Smith (1996), p. 148. Note that Smith's worry, at root, concerns the gap between physiological and semantic or intentional questions.

[2] Thus we read, for example, that computational approaches make possible "a science of structure and function divorced from material substance [that] ... can answer questions traditionally posed by psychologists" (Pylyshyn (1986, p. 68).

[3] See, for example, David Marr's (1982) distinction between the levels of computation, algorithm and implementation.

[4] See, for example, Simon (1982), Dawkins (1982), Clark (1997) Chapter 4.

the engineer's ahistorical perspective, the tinkerer's solution may look bizarre. Likewise, the processing strategies used by biological brains may surprise the computer scientist. For such strategies have themselves been evolved via a process of incremental, piecemeal, tinkering with older solutions.

More positively, biological evolution is *liberated* by being able to discover efficient but 'messy' or unobvious solutions: ones which may, for example, exploit environmental interactions and feedback loops so complex that they would quickly baffle a human engineer. Natural solutions (as we will later see) can exploit just about any mixture of neural, bodily and environmental resources along with their complex, looping and often non-linear interactions. Biological evolution is thus able to explore a very different solution space (wider in some dimensions, narrower in others) than that which beckons to conscious human reason.

There are, of course, ways around this apparent mismatch. Computationalists lately exploit so-called "genetic algorithms"[5] that roughly mimic the natural process of evolutionary search and allow the discovery of efficient but loopy and interactive adaptive strategies. Moreover, hard neuro-scientific data and evidence is increasingly available as a means of helping expand our imaginative horizons so as to better appreciate the way real biological systems solve complex problems. The one-time obsession with the "software level" is thus relaxing, in favour of an approach which grounds computational modelling in a more serious appreciation of the biological facts. In the next section, I highlight some challenging aspects of such recent research – aspects that will lead us directly to a confrontation with the situated brain.

Wetware and Some Robots

Recent work in cognitive neuroscience underlines the distance separating biological and 'engineered' problem solutions and displays an increasing awareness of the important interpenetration – in biological systems – of perception, thought and action. Some brief examples should help fix the flavour.

As a gentle entry point, consider some recent work on the neural control of monkey finger motions. Traditional wisdom depicted the monkey's fingers as individually controlled by neighbouring groups of spatially clustered neurons. According to this story, the neurons (in Motor Area 1, or M1) were organised as a 'somatotopic map' in which a dedicated neural sub-region governed each individual digit, with the sub-regions arranged in latero-medial sequence just like the fingers on each hand. This is a tidy, easily conceptualised solution to the problem of finger control. But it is the engineer's solution, not (it now seems) that of Nature.

---

[5] For a review, see Clark (1997) Chapter 5.

Marc Schieber and Lyndon Hibbard[6] have shown that individual digit movements are accompanied by activity spread pretty well throughout the M1 hand area, and that precise, single digit movements actually require *more* activity than some multi-digit whole hand actions (such as grasping an object). Such results are inconsistent with the hypothesis of digit-specific local neuronal groups. From a more evolutionary perspective, however, the rationale and design are less opaque. Schieber (1990, p. 444) conjectures that the basic case, from an evolutionary perspective, is the case of whole hand grasping motions (used to grab branches, to swing, to acquire fruits, etc.) and that the fundamental neural adaptations are thus geared to the use of simple commands which exploit inbuilt synergies[7] of muscle and tendon so as to yield such coordinated motions. The 'complex' coordinated case is thus evolutionarily basic and neurally atomic. The 'simple' task of controlling, for example,, an individual digit represents the harder problem and requires more neural activity, namely the use of some motor cortex neurons to *inhibit* the synergetic activity of the other digits. Precise single digit movements require the agent to tinker with whole-hand commands, modifying the basic synergetic dynamics (of mechanically linked tendons, etc.) adapted to the more common task.

Consider next a case of perceptual adaptation. The human perceptual system can, we know (given time and training), adapt in quite powerful ways to distorted or position-shifted inputs. For example, subjects can learn how to coordinate vision and action while wearing lenses that invert the entire visual scene so that the world initially appears upside down. After wearing such lenses for a few days, the world is seen to flip over – various aspects of the world now appear to the subject to be in the normal upright position. Remove the lenses and the scene is again inverted until readaptation occurs.[8] Thach et al. (1992) used a variant of such experiments to demonstrate the motor-specificity of some perceptual adaptations. Wearing lenses which shifted the scene *sideways* a little, subjects were asked to throw darts at a board. In this case, repeated practice led to successful adaptation,[9] but of a motor-loop specific kind. The compensation did not "carry over" to tasks involving the use of the non-dominant hand to throw, or to an underarm variant of the visual overarm throw. Instead, adaptation looked to be restricted to a quite specific combination of gaze angle and throwing angle: the one used in overarm, dominant-hand throwing.

Something of the neural mechanisms of such adaptation is now understood. It is known, for example, that the adaptation never occurs in patients with

[6] Schieber & Hibbard (1993).

[7] The notion of synergy aims to capture the idea of links that constrain the collective unfolding of a system comprising many parts. For example, the front wheels of a car exhibit a built-in synergy which allows a single driver 'command' (at the steering wheel) to affect them both at once. Synergetic links may also be learnt, as when we acquire an automated skill, and may be neurally as well as brute-physiologically grounded. See Kelso (1995), pp. 38, 52.

[8] For a survey of such experiments, see Welch (1978).

[9] In this case, *without* any perceived shift in the visual scene.

generalised cerebellar cortical atrophy, and that inferior olive hypertrophy leads to impaired adaptation. On the basis of this and other evidence, Thach et al. speculated that a learning system implicating the inferior olive and the cerebellum (linked via climbing fibres) is active both in prism adaptation and in the general learning of patterned responses to frequently encountered stimuli. The more general lesson, however, concerns the nature of the perception-action system itself. For it increasingly appears that the simple image of a general purpose perceptual system delivering input to a distinct and fully independent action system is biologically distortive. Instead, perceptual and action systems work together, in the context of specific tasks, so as to promote adaptive success. Perception and action, on this view, form a deeply inter-animated unity.

Further evidence for such a view comes from a variety of sources. Consider, for example, the fact that the primate visual system relies on processing strategies that are not strictly hierarchic but instead depends on a variety of top-to-bottom and side-to-side channels of influence. These complex inner pathways allow a combination of multiple types of information (high-level intentions, low-level perception and motor activity) to influence all stages of visual processing. The Macaque monkey (to take one well-studied example) possesses about 32 visual brain areas and over 300 connecting pathways. The connecting pathways go both upwards and downwards (e.g., from V1 to V2 and back again) and side-to-side (between subareas in V1) – see, for example,, Felleman and VanEssen (1991). Individual cells at 'higher' levels of processing, such as V4 (visual area 4) do, it is true, seem to specialise in the recognition of specific geometric forms. But they will each also respond, in some small degree, to many other stimuli. These small responses, spread unevenly across a whole population of cells, can carry significant information. The individual cells thus function not only as narrowly tuned single feature detectors but also as widely tuned filters reacting to a whole range of stimulus dimensions (see Van Essen & Gallant 9 (1994)). Moreover, the responses of such cells now look to be modifiable both by attention and by details of local task-specific context (Knierim & Van Essen (1992)).[10]

More generally, back projecting (corticocortical) connections tend, in the monkey, to outnumber forward ones,.that is, there are more pathways leading from deep inside the brain outwards towards the sensory peripheries than vice versa.[11] Visual processing may thus involve a variety of criss-crossing influences which could only roughly, if at all, be described as a neat progression through a lower-to-higher hierarchy.

Such complex connectivity opens up a wealth organisational possibilities in which multiple sources of information combine to support visually guided

---

[10] In a somewhat related vein, Caminiti et al. (1990) show that the directional preference of individual cells encoding reaching movement commands varies according to initial arm position. See also discussion in Jeannerod (1997), Chapter 1.

[11] Though much of the connectivity is reciprocal. See Van Essen & Anderson (1990), Churchland et al. (1991), p. 40.

action. Examples of such combinations are provided by Churchland, Ramachandran and Sejnowski (1994) who offer a neurophysiologically grounded account of what they term "interactive vision." The interactive vision paradigm is there contrasted with the approaches that assume a simple division of labour in which perceptual processing yields a rich, detailed inner representation of the visual scene which is later given as input to the reasoning and planning centres which in turn calculate a course of action and send commands to the motor effectors. This simple image (of what roboticists call a "sense-think-act" cycle) is, it now seems, not true to the natural facts. In particular:

1. daily agent-environment interactions often do not require the construction and use of detailed inner models of the full visual scene;
2. low-level perception may "call" motor routines that yield *better perceptual input* and hence improve information pick-up;
3. real-world actions may sometimes play an important role in the computational process itself and
4. the internal representation of worldly events and structures may be less like a passive data-structure or description and more like a direct recipe for action.

Evidence for proposition 1 comes from a series of experiments in which subjects watch images on a computer screen. Subjects are allowed to examine an on-screen pictorial display. Then, as they continue to saccade around the scene (focusing first on one area, then another) small changes are made to the currently unattended parts of the display. The changes are made during the visual saccades. It is an amazing fact that, for most of the part,[12] quite large changes go unnoticed: changes such as the replacement of a tree by a shrub, or the addition of a car, deletion of a hat, etc. Why do such gross alterations remain undetected? A compelling hypothesis is that the visual system is not even attempting to build a rich, detailed model of the current scene but is instead geared to using frequent saccades to retrieve information *as and when it is needed* for some specific problem-solving purpose. This fits nicely with Yarbus' classic (1967) finding that the pattern of such saccades varies (even with identical scenes) according to the type of task the subject has been set (e.g., to give the ages of the people in a picture, to guess the activity they have been engaged in, etc.). According to both Churchland et al. (1994) and Ballard (1991), we are prone to the illusion that we constantly command a rich inner representation of the current visual scene because we are able to perform fast saccades, retrieving information as and when required. (An analogy:[13] a modern store may present the illusion of having a massive amount of goods stocked on the premises, because it always has what you want when you want it. But modern computer ordering systems can automatically count off sales and requisition new items so

[12] The exception is if subjects are told in advance to watch out for changes to a certain feature. See McConkie (1990), Churchland et al. (1994).

[13] Thanks to David Clark for pointing this out.

that the necessary goods are available just when needed and barely a moment before. This fine-tuned ordering system offers a massive saving of on-site storage whilst tailoring supply directly to customer demand.)

Contemporary research in robotics avails itself of these same economies. One of the pioneers of "new robotics," Rodney Brooks (see e.g., Brooks, 1991) coined the slogan, "the world is its own best model" to capture just this flavour. A robot known as Herbert (Connell, 1989), to take just one example, was designed to collect soft drink cans left around a crowded laboratory. But instead of requiring powerful sensing capacities and detailed advance planning, Herbert got by (very successfully) using a collection of coarse sensors and simple, relatively independent, behavioural routines. Basic obstacle avoidance was controlled by a ring of ultrasonic sound sensors that brought the robot to a halt if an object was in front of it. General locomotion (randomly directed) was interrupted if Herbert's simple visual system detected a roughly table-like outline. At this point, a new routine kicks in and the table surface is swept using a laser. If the outline of a can is detected, the whole robot rotates until the can is centred in its field of vision. This simple physical action simplifies the pick-up procedure by creating a standard action-frame in which a robot arm, equipped with simple touch sensors, gently skims the table surface dead ahead. Once a can is encountered, it is grasped, collected and the robot moves on. Notice, then, that Herbert succeeds without using any conventional planning techniques and without creating and updating any detailed inner model of the environment. Herbert's 'world' is composed of undifferentiated obstacles and rough table-like and can-like outlines. Within this world the robot also exploits its own bodily actions (rotating the 'torso' to centre the can in its field of view) so as to greatly simplify the computational problem involved in eventually reaching for the can. Herbert is thus a simple example both of a system that succeeds using minimal 12 representational resources and one in which gross motor activity helps streamline a perceptual routine(as suggested in proposition (2) above).

The interactive vision framework envisages a more elaborate natural version of this same broad strategy, namely the use of a kind of perceptuo-motor loop whose role is to make the most of incoming perceptual information by combining multiple sources of information. The idea here is that perception is not a passive phenomenon in which motor activity is only initiated at the endpoint of a complex process in which the animal creates a detailed representation of the perceived scene. Instead, perception and action engage in a kind of incremental game of tag in which motor assembly begins long before sensory signals reach the top level. Thus, early perceptual processing may yield a kind of proto-analysis of the scene, enabling the creature to select actions (such as head and eye movements) whose role is to provide a slightly upgraded sensory signal. That signal may, in turn, yield a new proto-analysis indicating further visuo-motor action and so on. Even whole body motions may be deployed as part of this process of improving perceptual pick-up. Foveating an object can, for example, involve motion of the eyes, head, neck and torso. Churchland et al. (1994, p. 44) put it well: "watching Michael Jordan play basketball or a group of ravens steal a caribou corpse from a wolf tends to underscore the integrated, whole-body character of visuomotor coordination." This integrated character is consistent with neuro-physiological and neuro-anatomical data which show the influence of motor signals in visual processing. There are – to take just two small examples – neurons sensitive to eye position in V1, V3 and LGN (lateral geniculate nucleus), and cells in V1 and V2 that seem to know in advance about planned visual saccades (showing enhanced sensitivity to the target[14]).

Moving on to proposition (3) (that real-world actions may sometimes play an important role in the computational process itself), consider the task of distinguishing figure from ground (the rabbit from the field, or whatever). It turns out that this problem is greatly simplified using information obtained from head movement during eye fixation. Likewise, depth perception is greatly simplified using cues obtained by the observers own self-directed motion. As the observer moves, close objects will show more relative displacement than farther ones. That is probably why, as Churchland et al. (op.cit., p. 51) observe, head bobbing behaviour is frequently seen in animals: "a visual system that integrates across several glimpses to estimate depth has computational savings over one that tries to calculate depth from a single snapshot."

And so to proposition (4): that the neural representation of worldly events may be less like a passive data structure and more like a recipe for action. The driving force, once again, is computational economy. If the goal of perception and reason is to guide action (and it surely is, evolutionary speaking), it will often be simpler to represent the world in ways rather closely geared to the kinds of actions we want to perform. To take a simple example, an animal that uses its visual inputs to guide a specific kind of reaching behaviour (so as to acquire and ingest food) need not form an object-centred representation of the surrounding space. Instead, a systematic metrical transformation (achieved by a point-to-point mapping between two topographic maps) may transform the visual inputs directly into a recipe for reaching out and grabbing the food. In such a set-up, the animal does not need to do any computational work on an action-neutral inner model as to plan a reaching trajectory. The perceptual processing is instead tweaked, at an early stage, in a way dictated by the particular use to which the visual input is dedicated. This strategy is described in detail in P.M. Churchland's (1989, ch. 5) account of the "connectionist crab," in which research in Artificial Neural networks[15] is applied to the problem of creating efficient point-to-point linkages between deformed topographic maps.

In a related vein, Maja Mataric of the MIT Artificial Intelligence Laboratory has developed a neurobiologically inspired model of how rats navigate their

---

[14] See Churchland et al. (1994), p. 44, Wurz & Mohler (1976).

[15] For an accessible introduction, see P.M. Churchland (1995).

environments. This model exploits the kind of layered architecture[16] also used in the robot Herbert. Of most immediate interest, however, is the way the robot learns about its surroundings. As it moves around a simple maze, it detects landmarks which are registered as a combination of sensory input and current motion. A narrow corridor thus registers as a combination of forward motion and short lateral distance readings from sonar sensors. Later, if the robot is required to find its way back to a remembered location, it retrieves[17] an interlinked body of such combined sensory and motor readings. The stored 'map' of the environment is thus immediately fit to act as a recipe for action, since the motor signals are part of the stored knowledge. The relation between two locations is directly encoded as the set of motor signals that moved the robot from one to the other. The inner map is thus *itself* the recipe for the necessary motor actions. By contrast, a more classical approach would first generate a more objective map which would then need to be *reasoned over* in order to plan the route.

The Mataric robot (which is based on actual rat neurobiology – see McNaughton & Nadel (1990)) and the connectionist crab exemplify the attractions of what I call 'action-oriented representations' (Clark, 1997, p. 49): representations that describe the world by depicting it in terms of possible actions.[18] This image fits in nicely with several of the results reported earlier, including the work on monkey finger control and the motor loop specificity of 'perceptual' adaptation. The products of perceptual activity, it seems, are not always action-neutral descriptions of external reality. They may instead constitute direct recipes for acting and intervening. We thus glimpse something of the shape of what Churchland et al. (1994, p. 60) describe as a framework that is "motocentric"; rather than "visuocentric."

As a last nod in that same direction, consider the fascinating case of so-called "mirror neurons" (DiPellegrino et al. (1992)). These are neurons, in monkey ventral premotor cortex, that are action-oriented, context-dependent and implicated in both self-initiated activity and passive perception. They are active both when the monkey observes a specific action (such as someone grasping a food item) and when the monkey performs the same action, where sameness implies not mere grasping but the grasping of a food item (see also Rizzolatti et al. (1996)). The implication, according to the psychologist and neuroscientist Marc

---

[16] This is known as a 'subsumption' architecture, because the layers each constitute a complete behaviour-producing system and interact only in simple ways such as by one layers subsuming (turning off) the activity of another or by one layer's co-opting and hence 'building-in' the activity of another (see Brooks (1991)).

[17] By a process of spreading activation amongst landmark encoding nodes – see Mataric (1991).

[18] Such representations bear some resemblance to what the ecological psychologist J.J. Gibson called "affordances," although Gibson himself would reject our emphasis on inner states and encodings. For an affordance is the potential of use and activity that the local environment offers to a specific kind of being: chairs afford sitting (to humans) and so on. See Gibson (1979). The philosopher Ruth Millikan has developed a nice account of action-oriented representation under the label 'pushmipullyu representation' – see Millikan (1995).

Jeannerod is that "the action ... to be initiated is stored in terms of an action code, not a perceptual one" (Jeannerod (1997), p. 191).

Putting all this together suggests a much more integrated model of perception, cognition and action. Perception is itself tangled up with possibilities for action and is continuously influenced by cognitive contextual and motor factors. It need not yield a rich, detailed and action-neutral inner model awaiting the services of 'central cognition' so as to deduce appropriate actions. In fact, these old distinctions (between perception, cognition and action) may sometimes obscure, rather than illuminate, the true flow of effect. In a certain sense, the brain is revealed not as (primarily) the engine of reason or quiet deliberation, but as the organ of *environmentally situated control*. Action, not truth and deductive inference, are the key organising concepts. This perspective, however, begs to be taken a step further. For if brains are best understood as controllers of environmentally situated activity, then might it not be fruitful to locate the neural contribution as just one (important) element in a complex causal web spanning brains, bodies and world? This potential gestalt shift is the topic of the next section.

Wideware

Let us coin a term 'wideware' to refer to states, structures or processes that satisfy two conditions. First, the item in question must be in some intuitive sense environmental: it must not, at any rate, be realised within the biological brain or the central nervous system. Bodily aspects and motions, as well as truly external items such as notebooks and calculators, thus fit the bill. Second, the item (state, structure, process) must play a functional role as part of an *extended cognitive process*: a process geared to the promotion of adaptation success via the gathering and use of knowledge and information, and one that loops out in some non-trivial way, so as to include and exploit aspects of the local bodily and environmental setting.

Of course, even what is intuitively a fully internalised cognitive process will usually involve contact with the external world. That much is demanded by the very ideas of knowledge acquisition and information-gathering. The notion of wideware aims to pick out instead a somewhat more restricted range of cases – ones in which the *kind* of work that cognitive science has typically assigned to the inner workings of the brain is at least partly carried out by processes of storage, search and transformation realised using bodily actions and/or a variety of external media. Understanding the human mind, I shall argue, will require us to attend much more closely to the role of such bodily actions and external media than was once anticipated. To better fix the notion of wideware itself, consider first a very simple case involving 'bodily backdrop'.

Computer-controlled machines are sometimes used to fit small parts into one another. The error tolerance here is very low and sometimes the robot arm will fail to make a match. A computationally expensive solution exists in which the control system includes multiple feedback loops which signal such failures and prompt the machine to try again in a [17] minutely different orientation. It turns out, however, that a much cheaper, more robust and more efficient procedure is simply to mount the assembler arms on rubber joints that "give" along two spatial axes. This bodily backdrop allows the control device to dispense with the complex feedback loops. Thanks to the rubber mountings the parts "jiggle and slide into place just as if millions of tiny feedback adjustments to a rigid system were being continuously computed" (Mitchie & Johnson (1984), p. 95).

Mere bodily backdrop, however, does not really count as an instance of an extended cognitive process. The computational and information based operations are reduced, courtesy of the brute physical properties of the body. But they are not themselves extended *into* the world. Genuine cognitive and computational extension require, by contrast, that the external or bodily operations are themselves usefully seen as performing cognitive or information-processing operations. A simple phototropic (light following) robot – to take another negative example – does not constitute an extended cognitive or computational system. For although the presence of some external structure (light sources) is here vital to the robot's behavioural routines, those external structures are not doing the kind of work that cognitive science and psychology has typically assigned to inner neural activity. They are not acting so as to manipulate, store or modify the knowledge and information that the organism uses to reach its goals. Sometimes, however, external structures and bodily operations do seem to be the proper parts of the cognitive and computational processes themselves. Thus consider the use, in recent interactive vision research, of so-called *deictic pointers*. A classical (non-deictic) pointer is an inner state that can figure in computational routines but that can also "point" to further data-structures. Such pointing allows for both the retrieval of additional information and the binding of one memory location content to another. One use of classical pointers is thus to bind information about spatial location (e.g., "top left corner of visual field") to information about current features (e.g., "bright yellow mug"), yielding complex contents (in this case, "bright yellow mug in top left [18] corner of visual field" – see Pylyshyn (1989) for discussion). Since neural processing involves (more or less) distinct channels, for example, object properties, object location and object motion, binding is an important element in neural computation. But binding, it now seems, can sometimes be achieved by the use of actual bodily orientations instead of linking inner data structures. The story is complex, but the basic idea is straightforward. It is to set up a system so that bodily orientations (such as saccadic eye motions leading to object fixation) directly yield the kinds of benefits associated with classical binding. This Ballard et al. (in press) show how to use visual scene fixation to directly associate the features represented (the properties of the perceived object) with an external spatial location. Another example is the use so-called "do-it-where-I'm-looking" processing routines in which a bodily motion (e.g., grasping) is automatically directed to whatever location happens to be currently fixated in the visual field. In all these cases, the authors comment:

The external world is analogous to computer memory. When fixating a location the 19 neurons that are linked to the fovea refer to information computed from that location. Changing gaze is analogous to changing the memory reference in a silicon computer Ballard et al. (in press), p. 6.

Deictic binding provides a clear example of a case in which bodily motion does the *kind* of work that cognitive science has typically assigned only to inner neural activity. The deictic strategies use a combination of inner computation and gross bodily action to support a type of functionality once studied as a property of the inner neural system alone. Taking this perspective a step further, we next consider the use of external media as both additional memory and as potent symbol-manipulating arenas.

Portions of the external world, it is fairly obvious, often function as a kind of extra-neural memory store. We may deliberately leave a film on our desk to remind us to take it for developing. Or we may write a note 'develop film' on paper and leave that on our desk instead. As users of words and texts, we command an especially potent means of offloading data and ideas from the biological brain onto a variety of external media. This trick, I think, is not to be underestimated. For it affects not just the quantity of data at our command, but also the kinds of operation we can bring to bear on it. Words, texts, symbols and diagrams often figure *intimately* in the problem-solving routines developed by biological brains nurtured in language-rich environmental settings. Human brains, trained in a sea of words and text, will surely develop computational strategies that directly 'factor-in' the reliable presence of a variety of external props and aids. The inner operations will then complement, but not replicate, the special manipulative potentials provided by the external media.

Consider, for example,[19] the process of writing an academic paper. You work long and hard and at day's end you are happy. Being a good physicalist, you assume that all the credit for the final intellectual product belongs to your brain: the seat of human reason. But you are too generous by far. For what really happened was (perhaps) more like this. The brain supported some re-reading of old texts, materials and notes. Whilst re-reading these, it responded by generating a few fragmentary ideas and criticisms. These ideas and criticisms were then stored as more marks on paper, in margins, on computer discs, etc. The brain then played a role in re-organising this data, on clean sheets, adding new on-line reactions and ideas. The cycle of reading, responding and external re-organisation is repeated. Finally, there is a product. A story, argument or theory. But this intellectual product owes a lot to those repeated loops out into the environment. Credit belongs to the agent-in-the-world. The biological brain is just a part (albeit a crucial and special part) of a spatially and temporally extended process, involving lots of extra-neural operations, whose joint action

[19] This example is borrowed from Clark (1995).

creates the intellectual product. There is thus a real sense (or so I would argue) in which the notion of the 'problem-solving engine' is really the notion of the *whole caboodle*: the brain and body operating within an environmental setting.

Consider, by way of analogy, the idea of a swimming machine. In particular, consider the bluefish tuna.[20] The tuna is paradoxically talented. Physical examination suggests it should not be able to achieve the aquatic feats of which it is demonstrably capable. It is physically too weak (by about a factor of 7) to swim as fast as it does, to turn as compactly as it does, to move off with the acceleration it does, etc. The explanation (according to the fluid dynamicists, M & G Triantafyllou) is that these fish actively create and exploit additional sources of propulsion and control in their watery environments. For example, the tuna use naturally occurring eddies and vortices to gain speed, and they flap their tails so as to actively create additional vortices and pressure gradients which they then exploit for quick take-offs, etc. The real swimming machine, I suggest, is thus the fish *in its proper context*: the fish plus the surrounding structures and vortices than it actively creates and then maximally exploits. The *cognitive machine*, in the human case, looks similarly extended (see especially Dennett (1995), Chapter 11 and 12). We actively create and exploit multiple linguistic media, yielding a variety of contentful structures and manipulative opportunities whose reliable presence is then factored deep into our problem-solving strategies.

Implications

Software, wetware and wideware, if our story is to be believed, form a deeply inter-animated triad. The computational activities of the brain will be heavily sculpted by its biological 'implementation'. And there will be dense complementarity and cooperation between neural, bodily and environmental forces and factors. What the brain does will thus be precisely fitted to the range of complementary operations and opportunities provided by bodily structure, motion and the local environment. In the special case of human agency, this includes the humanly generated 'whirlpools and vortices' of external, symbol-laden media: the explosion of wideware made available by the ubiquitous devices of language, speech and text. The picture that emerges is undeniably complex. But what does it really mean both for our understanding of ourselves and for the practice of scientific inquiry about the mind?

Certain implications for our vision of ourselves are clear. We must abandon the image of ourselves as essentially disembodied reasoning engines. And we must do so not simply by insisting that the mental is fully determined by the

[20] The story is detailed in Triantafyllou & Triantafyllou (1995) and further discussed in Clark (1997). A 49 inch eight-segment anodised aluminum and lycra robot tuna is being used at MIT (Massachusetts Institute of Technology) to test the details of the theory.

physical, but by accepting that we are beings whose neural profiles are profoundly geared so as to press maximal benefit from the opportunities afforded by bodily structure, action and environmental surroundings. Biological brains are, at root, controllers of embodied action. Our cognitive profile is *essentially* the profile of an embodied and situated organism. Just how far we should then press this notion of cognitive extension, however, remains unclear. Should we just think of ourselves as cognitive agents who co-opt and exploit surrounding structures (e.g., pen and paper) so as to expand out problem-solving capacities, or is there a real sense in which the cognitive agent (as opposed to the bare biological organism) is thus revealed as an extended entity incorporating brain, body and some aspects of the local environment? Normal usage would seem to favour the former. But the more radical interpretation is not as implausible as it may initially appear.

Consider an example. Certain Alzheimer's sufferers maintain an unexpectedly high level of functioning within the normal community. These individuals should not – given their performance on a variety of standard tests – be capable of living as independently as they do. Their unusual success is explained only when they are observed in their normal home environments,[21] in which an array of external props and aids turn out to serve important cognitive functions. Such props and aids[22] may include the extensive use of labels (on rooms, objects, etc.), the use of a "memory book" containing annotated photos of friends and relatives, the use of a diary for tasks and events and simple tactics such as leaving all important objects (check book, etc.) in open view so as to aid retrieval when needed. The upshot, clearly, is an increased reliance on various forms of wideware (or 'cognitive scaffolding' – see Clark (1997)) as a means of counterbalancing[23] a neurally based deficit. But the pathological nature of the case is, in a sense, incidental. Imagine a whole community whose linguistic and cultural practices evolved so as to counterbalance a normal cognitive profile (within that community) identical to that of these individuals. The external props could there play the same functional role (of complementing a certain neural profile) but without any overtone of pathology-driven compensation. Finally, reflect that our own community is just like that. Our typical neural profile is different, to be sure, but *relative to that profile*, the battery of external props and aids (laptops, filofaxes, texts, compasses, maps, slide-rules, etc.) play just the same role. They offset cognitive limitations built into the basic

---

[21] See Edwards, Baum & Morrow-Howell (1994), Baum (1991).

[22] Some of this additional structure is maintained and provided by family and friends. (But similarly, much of our own wideware is provided by language, culture and institutions which we do not ourselves create.)

[23] Such counterbalancing is, as Marcel Kinsbourne has usefully reminded me, a somewhat delicate and complex matter. The mere provision of the various props and aids is useless unless the patient remains located in a stable, familiar environment. And the ability of different patients to make use of such added environmental structure itself varies according to the nature and extent of the neurally based deficit.

biological system. Now ask yourself what it would mean – in the case of the Alzheimer sufferer – to maliciously damage that web of external cognitive support? Such a crime has, as Daniel Dennett once remarked, much of the flavour of a harm to the *person*, rather than simple harm to property. The same may well be true in the normal case: deliberate theft of the poet's laptop is a very special kind of crime. Certain aspects of the external world, in short, may be so integral to our cognitive routines as to count as *part of the cognitive machinery* itself (just as the whorls and vortices are, in a sense, part of the swimming machinery of the Tuna). It is thus something of a question whether we should see the cogniser as the bare biological organism (that exploits all those external props and structures) or as the organism-plus-wideware. To adopt the latter perspective is to opt for a kind of "extended phenotype" view of the mind, in which the relation between the biological organism and the wideware is as important and intimate as that of the spider and the web.[24]

The implications for the specific study of the mind are, fortunately, rather less ambiguous. The vision of the brain as a controller of embodied and situated action suggests the need to develop new tools and techniques capable of investigating the brain (literally) in action: playing its part in problem-solving routines in (as far as possible) a normal ecological context. Very significant progress has already been made, of course. Non-invasive scanning techniques such as Positron Emission Topography (PET) and fast Magnetic Resonance Imaging (MRI) represent a giant leap beyond the use of single-cell recordings from anaesthetised animals. But we should not underestimate the distance that remains to be covered. For example, experimenters recently carried out some neural recordings from a locust in free flight.[25] This kind of ecologically realistic study is clearly mandated by the kinds of consideration we have put forward. Yet such investigations remain problematic due to sheer technological limitations. In the locust case, the researchers relied upon tiny radio transmitters implanted in the insect. But the information pick-up remained restricted to a scanty two channels at a time.

Moreover, it is not just the information-gathering techniques that need work, but also the analytic tools we bring to bear on the information gathered. If we take seriously the notion that brain, body and world are often united in an extended problem-solving web, we may need to develop analytic and explanatory strategies that better reflect and accommodate this dense inter-animation. To this end, there are (so far) two main proposals on the table. One proposal[26] is to use the tools of dynamical systems theory. This is a well established mathematical framework for studying the temporal evolution of states within complex systems. Such a framework appeals, in part, because it allows us to use a single mathematical formalism to describe both internal and external organisations and (hence) allows us to treat complex looping interactions (ones that

---

[24] See Dawkins (1982).

[25] See Kutsch et al. (1993). Thanks to Joe Faith for drawing this example to my attention.

[26] See, for example,, Thelen & Smith (1994), Port and Van Gelder (1995).

criss-cross brain, body and world) in a deeply unified manner. The other proposal[27] is to take the kinds of analysis we traditionally applied only to the inner states and extend it outwards. This means sticking with talk of representations and computations, but applying these ideas to extended organisations encompassing, for example, multiple individuals, maps, texts and social institutions. (My own view, which I will not defend here, is that we need to combine the two approaches by defining new, dynamical (process-based) ways of understanding key terms like "representation" and "computation".)[28]

Finally, the pervasive notion of a neural code or codes is now in need of a major overhaul. If the notion is to apply to real biological systems, it must be relieved of a good deal of excess baggage. Natural neural codes, as we saw in Section 2 (Wetware and some robots), will often be closely geared to the particular details of body and world. The coding for single finger motion is unexpectedly baroque, courtesy of the need to actually combat basic synergies created by the system of mechanically linked tendons. Other codes (e.g. , for whole hand grasping motions) prove unexpectedly simple and direct. Moreover, there is evidence of the widespread use of task-specific, motor-oriented and context-dependent encodings. Mirror neurons, recall, code for context-specific actions (e.g., grasping food) and function both in passive perceptions and in active grasping. And interactive vision routines press large benefits from minimal and often task-specific forms of internal encoding. In all these cases, we discern a much more austere and action-oriented vision of neural encoding – a vision stripped of the excess baggage of a single, rich, language-like inner code.[29]

The simple, almost common-sensical, notion of the brain as a system evolved so as to guide the actions of embodied agents in rich real-world settings thus yields substantial and sometimes challenging fruit. Gone is the vision of the environment as simply as a source of problem-specifying inputs and an arena for action. Instead, both basic and imposed aspects of environmental order and complexity now emerge as fundamental components of natural problem-solving behaviour. Gone too is the vision of the human brain as an organ of pure reason. In its place, we encounter the brain as a locus of action-oriented and activity-exploiting problem-solving techniques, and as a potent generator and exploiter of cognition-enhancing wideware. Taking this vision seriously, and turning it into a concrete and multi-disciplinary research program, presents an exciting new challenge for the sciences of the mind.

Acknowledgments I am very grateful to Stephen Graubard and the participants at the Daedalus Authors Meeting (Paris, October 1997) for a wealth of useful advice, good criticism and wise counsel. Special thanks to Jean Pierre Changeux, Marcel Kinsbourne, Vernon Mountcastle, Guilio Tonini, Steven Quartz and Semir Zeki. As usual, any remaining errors are all my own.

---

[27] See, for example,, Hutchins (1995).

[28] See Clark (1997), Crutchfield & Mitchell (1995), Mitchell et al. (1994), van Gelder (1995).

[29] For example,, Fodor (1975), Fodor & Pylyshyn (1988).

References

Ballard, D. (1991). Animate vision. *Artificial Intelligence, 48*, 57–86.

Ballard, D., Hayhoe, M., Pook, P. & Rao, R. (1997). Deictic codes for the embodiment of cognition. *Behavioral & Brain Sciences 20*, 723–767.

Baum, C. (1991). Addressing the needs of the cognitively impaired elderly from a family policy perspective. *American Journal of Occupational Therapy, 45*(7), 594–606.

Brooks, R. (1991). Intelligence without representation. *Artificial Intelligence, 47*, 139–159.

Caminiti, R., Johnson, P., Burnod, Y., Galli, C., & Ferrina, S. (1990). Shift of preferred directions of premotor cortical cells with arm movements performed across the workspace. *Experimental Brain Research, 82*, 228–232.

Churchland, P., Ramachandran, V., & Sejnowski, T. (1994). A critique of pure vision. In C. Koch & J. Davis (Eds.), *Large-Scale Neuronal Theories of the Brain* . Cambridge, MA: MIT Press.

Churchland, P. M. (1989). *The Neurocomputational Perspective.* Cambridge: MIT/Bradford Books.

Churchland, P. M. (1995). *The Engine of Reason, the feat of the soul.* MIT press, Cambridge, MA.

Clark, A. (1995). I am John's brain. *Journal of Consciousness Studies, 2*(2), 144–148.

Clark, A. (1997). *Being There: Putting Brain, Body and World Together Again.* Cambridge, MA: MIT Press.

Clark, A. (1998). Twisted tales: Cansal Complexity and cognitive scientific explanation. *Minds & Machines. 8*, 79–99.

Connell, J. (1989). A Colony Architecture for an Artificial Creature. *MIT AI Lab Tech Report II, 5, 1.*

Crutchfield, J. & Mitchell, M. (1995). The evolution of emergent computation. *Proceedings of the National Academy of Sciences, 92*, 10742–10746.

Dawkins, R. (1982). *The Extended Phenotype.* Oxford: Oxford University Press.

Dennett, D. (1995). *Darwin's Dangerous Idea.* New York: Simon & Schuster.

DiPelligrino, J., Klatzky, R., & McCloskey, B. (1992). Time course of preshaping for functional responses to objects. *Journal of Motor Behavior, 21*, 307–316.

Edwards, D., Baum, C., & Morrow-Howell, N. (1994). Home environments of inner city elderly: Do they facilitate or inhibit function? *The Gerontologist: Program Abstracts of 47th Annual Scientific Meeting, 34*(1), 64.

Felleman, D., & VanEssen, D. (1991). Distributed hierarchial processing in primate visual cortex. *Cerebral Cortex, 1*, 1–47.

Fodor, J. (1975). *The Language of Thought.* New York: Crowell.

Fodor, J., & Pylyshyn, Z. (1988). Connectionism and cognitive architecture: A critical analysis. *Cognition, 28*, 3–71.

Gibson, J. J. (1979). *The Ecological Approach to Visual Perception.* Boston: Houghton-Mifflin.

Hutchins, E. (1995). *Cognition in the Wild.* Cambridge, MA: MIT Press.

Jacob, F. (1977). Evolution and Tinkering. *Science, 196*, 1161–1166.

Jeannerod (1997). *The Cognitive Neuroscience of Action.* Oxford: Blackwell.

Kelso, S. (1995). *Dynamic Patterns.* Cambridge: MIT Press.

Knierim, J., & Van Essen, D. (1992). Visual cortex: Cartography, connectivity and concurrent processing. *Current Opinion in Neurobiology, 2*, 150–155. 31

Kutsch, W., Schaarz, G., Fischer, H., & Kautz, H. (1993). Wireless transmission of muscle potentials during free flight of a locust. *Journal of Experimental Biology, 185*, 367–373.

Marr, D. (1982). *Vision.* San Francisco, CA: W.H. Freeman.

Mataric, M. (1991). Navigating with a Rat Brain: A neurobiologically inspired model for robot spatial representation. In J.A. Meyer & S. Wilson (eds) *From Animals to Animats I.* Cambridge: MIT Press.

McConkie, G. W. (1990). *Where vision and cognition meet.* Paper presented at the H.F.S.P. Workshop on Object and Scene Perception, Leuven, Belgium.
McNaughton, B., & Nadel, L. (1990). Hebb-Marr networks and the neurobiological representation of action in space. In M. Gluck & D. Rumelhart (Eds.), *Neuroscience and Connectionist Theory* (pp. 1–64). Hillsdale, NJ: Erlbaum Associates.
Millikan, R. (1995). Pushmi-Pullyu representations. In J. Tomberline (ed) *Philosophical Perspectives 9: Connectionism and Philosophical Psychology*. Ridgeview, CA.
Mitchell, M. Crutchfield, J., & Hraber, P. (1994). Evolving cellular automata to perform computations. *Physica D, 75*, 361–391.
Mitchie, D., & Johnson, R. (1984). *The Creative Computer*. U.K.: Penguin.
Pylyshyn, Z. (Ed.). (1987). *The Robots Dilemma: The Frame Problem in Artificial Intelligence* . Norwood: Ablex.
Reeke, G., & Edelman, G. (1988). Real brains and artificial intelligence. *Daedalus, Winter*, 143–173.
Rizzolatti, G., Fadiga, L., & Fogassi, L. (1996). Premotor cortex and the recognition of motor actions. *Cognitive Brain Research, 3*, 131–141.
Schieber, M. (1990). How might the motor cortex individuate movements? *Trends in Neuroscience, 13*(11), 440–444. 32
Schieber, M., & Hibbard, L. (1993). How somatotopic is the motor cortex hand area? *Science, 261*, 489–492.
Simon, H. (1982). *Models of Bounded Rationality* (Vol. I and II). Cambridge, MA: MIT Press.
Smith, B. C. (1996). *On the Origin of Objects.* Cambridge, MA: MIT Press.
Thach, W., Goodkin, H., & Keating, J. (1992). The cerebellum and the adaptive coordination of movement. *Animal Review of Neuroscience, 15*, 403–442.
Thelen, E., & Smith, L. (1994). *A Dynamic Systems Approach to the Development of Cognition and Action.* Cambridge, MA: MIT Press.
Triantafyllou, M., & Triantafyllou, G. (1995). An efficient swimming machine. *Scientific American, 272*(3), 64–71.
Van Essen, D., & Anderson, C. (1990). Information processing strategies and pathways in the primate retina and visual cortex. In S. Zornetzer, J. Davis, & C. Lau (eds) *An Introduction to Neural and Electronic Networks* (p. 43–72). New York: Academic Press.
Van Essen, D., & Gallant, J. (1994). Neural mechanisms of form and motion processing in the primate visual system. *Neuron, 13*, 1–10.
van Gelder, T., & Port, R. (1995). It's about time: An overview of the dynamical approach to cognition. In R. Port & T. v. Gelder (Eds.), *Mind as Motion: Explorations in the Dynamics of Cognition* (pp. 1–44). Cambridge, MA: MIT Press.
Welch, R. (1978). *Perceptual Modification: Adapting to Altered Sensory Environments.* New York: Academic Press.
Wurz, R., & Mohler, C. (1976). Enhancement of visual response in monkey striate cortex and frontal eye fields. *Journal of Neurophysiology, 39*, 766–772.
Yarbus, A. (1967). *Eye movements and vision.* New York: Plenum Press.

Chapter 2
At the Potter's Wheel: An Argument *for* Material Agency

Lambros Malafouris

1

Consider a potter throwing a vessel on the wheel (Fig. 2.1). Think of the complex ways brain, body, wheel and clay relate and interact with one another throughout the different stages of this activity and try to imagine some of the resources (physical, mental or biological) needed for the enaction of this creative process. Focus, for instance, on the first minutes of action when the potter attempts to centre the lump of clay on the wheel. The hands are grasping the clay. The fingers, bent slightly following the surface curvature, sense the clay and exchange vital tactile information necessary for a number of crucial decisions that are about to follow in the next few seconds. What is it that guides the dextrous positioning of the potter's hands and decides upon the precise amount of forward or downward pressure necessary for centring a lump of clay on the wheel? How do the potter's fingers come to know the precise force of the appropriate grip? What makes these questions even more fascinating is the ease by which the phenomena which they describe are accomplished. Yet underlying the effortless manner in which the potter's hand reaches for and gradually shapes the wet clay lies a whole set of conceptual challenges to some of our most deeply entrenched assumptions about what it means to be a human agent.

There are two obvious ways to proceed in order to meet these challenges and answer these questions: the first is to turn and ask the potter directly. As a great deal of cross-cultural ethnographic observation will testify, confronted with the 'how do you do it?' question, potters would prefer to 'show you' rather than simply 'tell you' their answer. If, however, the question gets very precise, for instance, 'how did you decide the force of the grip?' or 'how did you decide the appropriate speed of the wheel' or 'when and how much water to add on the clay?', they usually have very little to say. They can do it but they do not know how they do it or they simply lack the means to express or communicate this

L. Malafouris
McDonald Institute for Archaeological Research, Cambridge University, Cambridge, UK
e-mail: Lm243@cam.ac.uk

C. Knappett, L. Malafouris (eds.), *Material Agency*,
DOI: 10.1007/978-0-387-74711-8_2, © Springer Science+Business Media, LLC 2008

Fig. 2.1 At the potter's wheel

form of tacit knowledge. No one – not even the potter himself – can have access to this type of information because no one – not even the potter himself – can tell the fingers how hard they can press the clay in and up so that the walls of the vessel will not collapse. When it comes to embodied skill, potters are no exception to the rules of action and material engagement. Potters know more than what they can tell or explain and their hands often have reasons of which their mind is not aware and which the clay may resist or accommodate. Verbal description, however detailed, can hardly capture the phenomenological perturbations of real activity and the reciprocality between the crafted and the crafter. This is also why the affordances of the wheel throwing technique need to be discovered each time, in real time and space, within the totality of the interactive parameters.

Let us now turn to the second way of answering our previous questions, namely to look for some 'internal' mental and inaccessible mechanism. From such a perspective, the potter's fingers do nothing but execute the orders of the potter's brain and it is there that we should be looking for an answer. The potter's fingers simply receive information from the clay and transmit it to the appropriate area inside the potter's brain; they have nothing to do with the central 'executive' mechanism responsible for the 'executive processing' and decision making. The moment you subscribe to the above popular scenario, you have already committed yourself also to a specific agency judgement. That is, you have already implicitly answered another question, what in this chapter I shall be calling the 'agency question', ie, who did it? Who is the author of the act? The paradox is that although the potter may again be totally unaware about how or when his brain is making all these fine small decisions or even about what precisely they consist of, this time, he is, more often than not, going to answer that question, with the ease of a natural-born dualist: 'I' did it. The following example from G. Bateson nicely illustrates this anthropocentric 'I did it-stance' that I shall be calling in this chapter the 'agency problem':

> Consider a man felling a tree with an axe. Each stroke of the axe is modified or corrected, according to the shape of the cut face of the tree left by the previous stroke. This self-corrective (i.e., mental) process is brought about by a total system, trees-eyes-brain-muscles-axe-stroke-tree; and it is this total system that has the characteristics of immanent mind. . .But this is *not* how the average Occidental sees the event sequence of tree felling. He says, "I cut down the tree" and he even believes that there is a delimited agent, the "self", which performed a delimited "purposive" action upon a delimited object (Bateson 1973, 318).

But what is this agency problem really about? Subject to the level of analysis (micro-macro), the agency problem can take many different forms. However, what hold those different forms together are two categorical errors that they have in common: The first is an error of *apparent mental causation* and the second and correlated one is that of *agency attribution*. According to Wegner, both errors pertain to the fact that people tend to experience conscious will, and thus agency, quite independently of any actual causal connection between their thoughts and actions (Wegner 2004, 654). The following example can take us to the heart of the issue:

> Imagine for a moment that by some magical process, you could always know when a particular tree branch would move in the wind. Just before it moved, you knew it was going to move, in which direction, and just how it would do it. Not only would you know this, but let us assume that the same magic would guarantee that you would happen to be thinking about the branch just before each move. You would look over, and then just as you realized it was going to move, it would do it! In this imaginary situation, you could eventually come to think that you were somehow causing the movement. You would seem to be the source of the distant branch's action, the agent that wills it to move. The feeling that one is moving the tree branch surfaces in the same way that one would get the sense of performing any action at a distance (Wegner 2004, 654).

The above example embodies the crux of Wegner's famous 'illusion of conscious will argument' (Wegner 2003; 2002) which directly relate to the crucial questions about 'what is the origin of an event we need to explain?' (see Law, this volume) and about 'who is the author of an act'? However, I should clarify that despite using Wegner's example as my starting point to the agency problem, my strategy for tackling this problem and my interpretation of the reasons behind it would be rather different and to a large extent contradictory to Wegner's account. In particular, following the Material Engagement approach

(Malafouris 2004), I will suggest that the agency problem is not so much the product of human illusion or some other attribution error of our left hemisphere 'interpreter' (Gazzaniga 1998) but of a certain acquired imbalance between mental and physical causality that destabilises the human cognitive equation.

To redress this imbalance at the root of the agency problem in this chapter I shall be introducing the notion of *material agency*. The concept itself, that is, *material agency*, is to some extent a misnomer, yet I believe it serves well my basic hypothesis which can be very simply expressed as follows: If human agency *is* then material agency *is*, there is no way that human and material agency can be disentangled. Or else, *while agency and intentionality may not be properties of things, they are not properties of humans either: they are the properties of material engagement, that is, of the grey zone where brain, body and culture conflate.*

To explore my working hypothesis and develop the argument for material agency, I shall be looking in between, rather than within, persons and things. Specifically, I shall be focusing on the brain-artefact interface (BAI) and using the potter's wheel as a good illustration of such a bio-interface. Besides my ethnographic and experiential familiarity with the task domain, there is an additional, perhaps even more important, reason behind my choice of the potting process as the focus of my discussion: I consider pottery making as a prototypical exemplar and one of the best and diachronic models of the active mind. Not only do I see the ways of potmaking as ways of thinking but I also believe that one can find few other diachronic and cross-cultural examples where all major ingredients of the human cognitive recipe are brought forth and actualised in such an explicit and to a large extent empirically accessible manner. Specifically, for the Material Engagement approach to the study of mind the potter's wheel is as the thermostat is to cybernetics or the computer is to computationalism. Moreover, and in addition to seeing at the continuum of potter's brain-body-clay-wheel what others are seeing in a Turing machine or a centrifugal governor, I also consider clay to be one of the earliest truly neuro-compatible materials in the history of humanity. Neuro-compatible here refers to materials that afford the flow of noetic activity beyond skin and skull thus bridging neural and cultural plasticity (Malafouris in press; Malafouris & Renfrew, in press). It is this flow that enables the hand of the potter, as I will argue below, to navigate upon the surface of clay with a minimal need of storage and internal processing. It is this meeting which, with a little help from 'active externalism' (Clark 1997; Clark & Chalmers 1998), can transform a prehistoric potsherd from a mute inert piece of matter to an index and constitutive residual component of the prehistoric mind (Malafouris in press).

2

But let me return to our 'agency problem' and see how this can be reformulated if thrown on the wheel. Once you look inside the dynamics of mediated action (Wertsch, 1998), a number of interesting questions can be raised about agency in (pottery-) making. I start with the one I consider the most basic of all: Let us say that the chosen clay was too porous, resulting in a vessel of low quality or causing the pot to crack during drying or explode during firing; who is to blame? Do not think of a scenario where no good quality clay is available but rather think of clay preparation as a 'technical choice'. The term 'technical choice' is used in archaeology and anthropology to describe the activity chain in material procurement and manufacture and by employing the term 'choice', we presume alternatives in this sequence that did not get chosen (Van der Leeuw 1993, 241; Schiffer & Skibo 1997, 29). So, one way to approach our previous question is to ask who or what is responsible for those choices? At first sight, it may appear that it is the potter who made those choices, but a closer look will reveal that, for example, the causal link between the crack, the choice of clay and the potter who made that choice is not as direct and straightforward as we might initially think. And if we accept that agency is essentially about doing and that the problem of agency is essentially about who or what is the cause of the doing, then what we need to try first to understand is the relation between agency and causality.

To this end it is necessary first to clarify an important distinction between the *sense of agency* and the *sense of ownership* (Gallagher 2000, 2005; Tsakiris & Haggard 2005). By *sense of agency* we refer to the potter's feeling that it is he who is moving his hands spreading out, pounding and shaping the clay. By the *sense of ownership* we refer to the potter's feeling that it is his hand that is moving. Two important points need to be underlined here: (a) The first point is that although our sense of agency and ownership are usually closely associated this does not necessarily have to be the case all the time. For example, although an experienced potter immersed in the shaping of a vessel will very often report that the sense of ownership, the sense that it is his hands that touch and move the clay, is experienced throughout the activity, the sense of agency, on the other hand, the feeling that it is he that is causing the movement, is very often disrupted. (b) The second point is that we are speaking about a *sense* and not about agency or ownership per se. That means that we may well have a very real sense of agency or ownership without in reality owning or causing our act whatsoever. It is one thing to say that only humans have a *sense* of agency, that is, the ability to refer to oneself as the author of one's own actions; it is another thing to say that only humans are agents in the sense of being able to initiate casual events with intentional character. I shall be returning to the intentional character of human experience in a later section; for now I want to yours upon the issue of causality. Whatever sense the potter is or is not having the question to be answered remains 'who' or 'what' is causing the act, or more specifically, the making of the pot.

Attempting to answer that by taking agency as a fixed human property it is to take as the starting point of analysis what should have been its end. The only available starting and obligatory point of passage for the emergence and determination of agency is that of material engagement. First the hand grasps the clay in the way the clay affords to be grasped, then the action becomes skill,

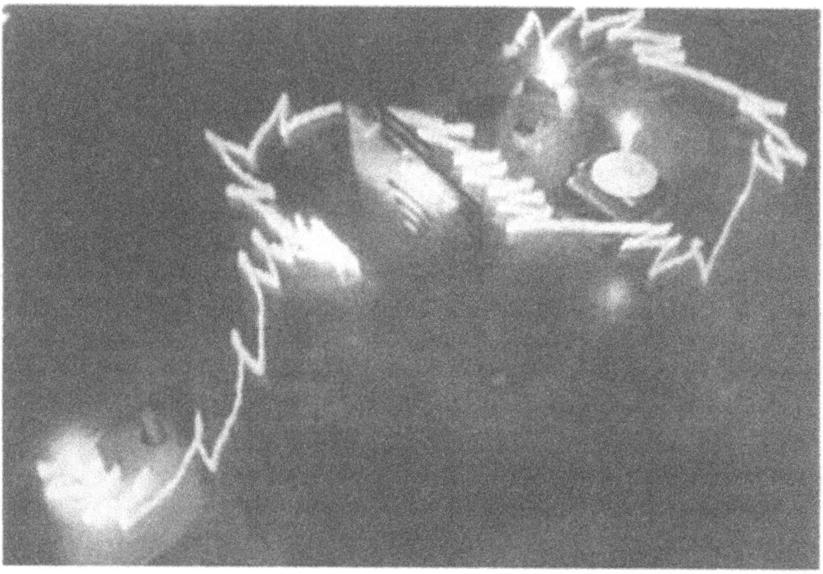

Fig. 2.2 Elsie searching for light (Candles were fixed to the turtles' shells and long exposures were used. The light streaks show the path of the turtle, © Burden Neurological Institute)

skill effects results and from those results that matter agency emerges. As I also discuss elsewhere (Malafouris 2004, see also Knappett 2006) the potter and the task-environment display a dynamic coupling between mind and matter that looks like a dance of agency not dissimilar to the one performed by Walter's 'turtles'[1] (Fig. 2.2).

In fact pushing my "tortoise" analogy further I suggest that it is a similar cybernetic transgression of the mind-body divide, like the one effected by

[1] I am referring to Walter's (1953) creation of the first autonomous robotic devices (*machina speculatrix*) baptised as Elsie and Elmer (for ElectroMechanical Robot, Light-Sensitive) later nicknamed after an "Alice in Wonderland" character as "tortoise". The devices though primitive from a mechanical and electronic point of view were capable of displaying unusual and unexpected forms of complex behaviour in the absence of any representational content. On the basis of their primitive circuitry, the tortoises were in a way structured to perform only two actions: (a) to avoid obstacles, retreating when they hit one and (b) to seek a light source. However, engaged with the environment, they were capable to produce emergent properties and in some cases what appeared as meaningful behaviour that could not be determined by their system components. Primitive as they might seem in the light of recent developments in the domain of AI the tortoise's managed to effect in practice a cybernetic transgression of the mind-body-world divide, materially exemplifying an embodied, performative cognitive system, one in which the mind-body-world components are continuous and equally necessary, with none hierarchically controlling the others. A premise that resonates well and to some extent anticipates the perspective of embodied and situated cognition as developed the last two decades in the works of Clark (1997), Hutchins (1995), Brooks (1991) and Van Gelder (1995).

Walter's turtles, that we see exemplified in the case of pottery making. Obviously to reduce the potter to some sort of human 'tortoise' in the above analogy is to leave out much. Nonetheless, we should bear in mind that the question that concern us here is not whether what is left out is important for what it means to be human but whether what is left out is of real importance for what it means to be an agent and my answer is that it is not. So far as agency is concerned the important thing to underline, and this is what our analogy does, is that this dance is between equal partners. This equality, symmetry or whatever word one chooses to express the relationship between the potter and the clay does not imply either that there are no important differences between the potter and the clay or that one of the two partners is not at times leading the dance. What it does very simply imply is that trying to separate cause from effect inside the loop of pottery making is like trying to construct a pot keeping your hands clean from the mud.

At the same time, simply to adopt an interactive perspective to action by recognising that actions seem to arise as a consequence of triggering or cueing by the external environment, in our case, the clay is not to say much really. Neither does interaction in itself deny that the actions are driven from within. Although few of us can resist the allure of a good phenomeno logical description that pull us inside this seamless flow of activity and agency, when we cut the flow and press the question of agency our inner Cartesian self wakes up to take control of the situation. It might well be that a part of the thinking takes place inside the head, a part of it in the body, part of it in the surrounding environment and the affordances of the tool-kit available to the potter, but at the end of the day is not the potter the one who really decides and intends what sort of vessel to produce? When it comes to the 'accountability question', it is the human side that makes the vital choices and takes the important decisions. For sure, many external factors (e.g., the texture of clay and its physical properties and may be even chemical consistency) may be allowed to determine some parts of the action but final responsibility rests with the potter. It is he who is to blame, a price that most people are willing to pay for the sake of free will or the 'illusion' of it (Wegner 2002; 2004).

Is there any way out of this? One way to proceed, I suggest, is to try and cut deep across the scales of time. In other words, try to develop a detailed temporal anatomy of the act. What is really important in this context is that our account of the causal hierarchy of events will not trivialise the complexities of the processes engaged in decision-making. To accomplish that, the starting point cannot be agency –the natural property of the human actor – the starting point should be time. More specifically the first condition of agency identification should be to define the portion of time which encapsulates the event you want to describe. Then follows the second criterion, which is deciding whether this portion of time constitutes a meaningful event in the larger enchainment of events that constitute the activity you seek to explain. To treat agency as the natural atemporal property of human beings is to strip the notion of agency of any analytic value and significance.

The importance of the temporal element is crucial here in more than one level that it might be useful to distinguish. First, we have a temporal relation between the enacted elements (biological or cultural). Some of these elements, for example, the hand, the clay, the wheel are in constant, permanent interaction from the beginning to the end of the process. They are the constitutive, one might say 'universal' ingredients, of the act of pottery-making. For example, the activities of *squeezing, supporting* and *controlling* the shape of the vessel while it is plastic and *turning* the pot in the hands, can be considered as the 'executive functions' of pottery making (Van der Leeuw 1994, 137). Other elements or actions, however, are recruited at different points and have a more transient role (e.g., cutting, scraping or smoothing). This does not mean that their role is not important, it simply means that the type of relation (transient or permanent) needs to be clarified if we are to decide which events matter and are meaningful for the act.

Thus, in what follows, I shall attempt to construct a *chrono-architecture* of the act. Such a *chrono-architecture* would be critical for understanding how the agency attribution is made, that is, how an action is attributed to its proper origin and how this origin might not be identical with the subject's *conscious agency judgement*.

3

Let us return to where we started this chapter, the potter throwing the vessel. Think of the moment before the act, the moment where the intention to act is formed. To give our thought experiment a Wittgenstein (1953) twist let us ask the following: When the potter moves his arm reaching towards the clay, what is left after subtracting the fact that his arm is moved? This simple and at the same time immensely complicated question has received recently a very interesting and much contested answer. Some years ago, Benjamin Libet through a series of experimental studies (1983; 1985; 1999) discovered for the first time a possible neural precursor (the so-called 'readiness potential', RP) to conscious intent to act (Fig. 2.3). More specifically Libet's claim was that what

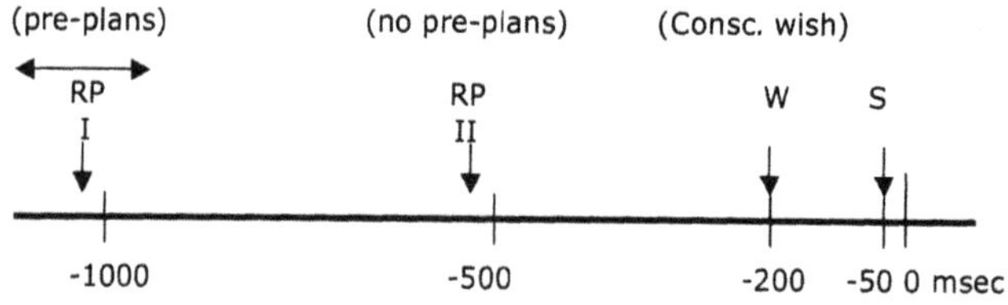

Fig. 2.3 Libet's diagram of sequence of events that precedes a self-initiated voluntary act. Time 0 refers to the time the electromyogram (EMG) detects muscle activation, (RP) stands for readiness potential which indicates neural activity. (W) indicates subjective awareness to move which appear at about 200 msec before the act (recorded in EMG) but about 350 msec after the first neuro-physiological indication of pre-planning (RPII) at about –550 msec (from Libet 1983 et al.)

I may call the micro *chrono-architecture* of a given voluntary act can be described in three stages: stage (1) a specific electrical change in the brain (the RP) that begins 550 ms before the act, stage (2) human subjects became aware of the intention to act 350–400 msec after RP starts and finally, stage (3) human subjects became aware of the intention 200 msec before the motor act.

In Libet's words "The brain 'decides' to initiate or, at least, to prepare to initiate the act before there is any reportable subjective awareness that such a decision has taken place" (1985, 536). This means that conscious will cannot be the true agent. Conscious will, according to the recordings of Libet, clearly appears after the RP. The potter's brain prepares to shape the clay before the appearance of a conscious urge or intention to do so. How might this be possible? The conventional neuro-scientific answer would be that the brain anticipates or predicts that act before the potter becomes consciously aware of it. The feeling of being in control of ones bodily movement can be explained in terms of the complex way the brain predicts movement (Howhy & Frith 2004).

But if the will of the potter to move his arm is not what initiates and really causes the act; then what is it? What is that tells the potter's brain to carry out a given activity? Moreover, what about human free will? Are we determined by deeply subconscious physiological process after all? The answer is offered by Libet himself: it might be that the 'readiness potential' (RP) precedes the appearance of the subject's awareness of the conscious wish to act by at least 350 msec but the conscious wish to act also precedes the final motor act by about 150–200 msec (Libet 1985). In other words, conscious will could still block or 'veto' the act; human agency and free will is saved once more. Yet not without a price because if the conscious veto act is itself preceded by some unconscious processes or veto 'readiness potentials' then we are back from where we started.

So, to put it simply, is the potter's brain to blame? And if it is not to be blamed, as I intend to argue in this chapter, then how may one proceed to resolve the agency question and avoid committing some form of the usual homunculus fallacy?

So far in this chapter, I have tried to show that the answer to the above questions may not be as obvious as it might seem. But even if we recognise the potting process as a distributed assembly bound by synchronisation of neurons, fingers and clay, it will only get us so far, at least where agency is concerned. There is still plenty of room left to accommodate all the usual agency-attribution errors and,thus, for ascribing agency solely to the side of the human. Yes, one may well accept that part of the action knowledge is embodied in the affordances of the wheel, but is it not the foot of the potter which set the wheel in motion? Is it not the potter's hand actualising the creative potential of this technology? Is it not because of the potter's intention that all those technical choices that affect vessel shape and size came into being in the first place? Is it not the potter's desires and doings that initiate the chain of events?

Strange as it might seem, *mediation* has it that the answer to the above questions can be both 'Yes' and 'No'. 'Yes', as we say 'hand-made', because it is certainly the potter's hand shaping the clay. 'No', as we say 'wheel-made', because whether we like it or not, introducing the wheel a very powerful 'dynamic attractor' shapes the field of action and has a share and saying on our will and intentions.

Clearly more is needed in order to disambiguate the situation. Thus I will now turn my focus to another notion with a crucial bearing on the problem of agency, that is, intentionality.

4

There is no doubt that intentionality, known also as 'object-directedness' is very often perceived as the major diagnostic feature of agency and as such presents a significant obstacle for any discussion of material agency in a proper sense. In contemporary philosophy of mind, it is usually seen as a fundamental property of human mental states to be "directed at, or about, or of objects and states of affairs in the world" (Searle 1983, 1; cf. also Dennet 1987). Seen from this 'internalist' philosophical perspective, the issue of intentionality appears to be fairly straightforward – no room for active externalism here. Intentional states are essentially projections that aim at, point at and extend toward objects or representations. They are *of* or *about* things, whereas no physical phenomena are in themselves *of* or *about* anything. As such, it appears initially that if we accept a close correlation between intentionality and agency, we have no option but to admit that as long as the former is conceived as being strictly a human property, so it must be the case also for the latter. In other words, if the nature of agency is intentional then it has to be a human property; things cannot exhibit intentional states. Indeed, the orthodox view, as Gell describes it, defines the agent according to the "capacity to initiate causal events in his/her vicinity, which cannot be ascribed to the current state of the physical cosmos, but only to a special category of mental states; that is, intentions" (Gell 1998, 19).

However, we have shown in our previous discussion that none of the above claims necessarily follows – at least not in all cases. Without denying that agency and intentionality are intimately connected, I believe that our understanding of this relation is based on a misunderstanding of the issues involved and as such it needs to be placed on a new footing. Thus in the following, I will attempt to dissociate agency from intentionality. My principal means to do so is by clarifying first the important difference between *prior intention* and *intention in action* drawing upon the work of John Searle.

What is an action? For Searle, the meaning of action is that of "a causal and intentional transaction between mind and the world" (1983, 88). More specifically, he describes activity as being composed of two essential parts: An intentional state in the mind and an external movement in the world. Based on that assumption Searle differentiates between two types of intentional states (Fig. 2.4):

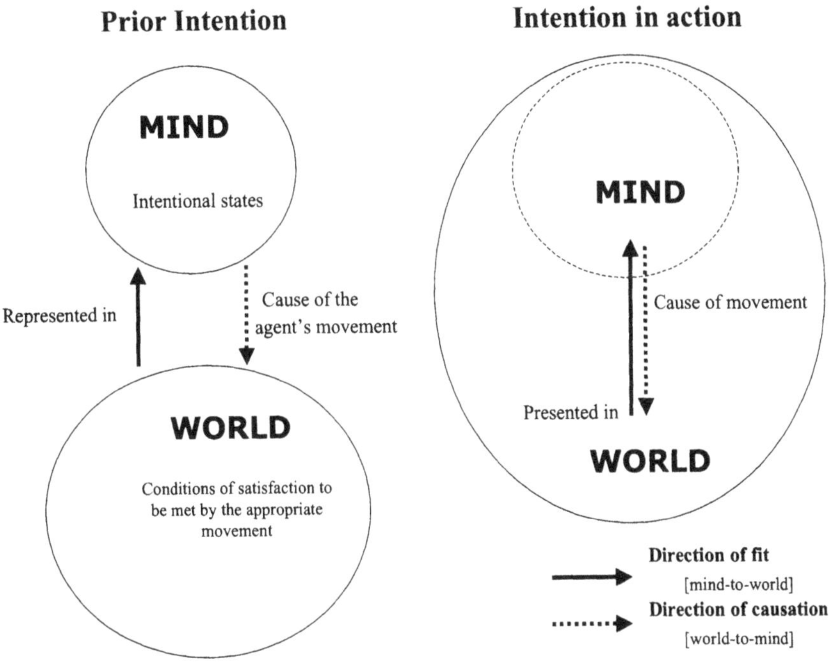

Fig. 2.4 'Prior intention' and 'intention in action'

The first type is called "prior intention" and is referring to premeditated or deliberate action where the intention to act is presumably formed in advance of the action itself. The second type of intentionality is called "intention-in-action" and is referring to non-deliberate everyday activity where no intentional state can be argued as being formed in advance of the action itself.

Moreover, Searle analyses intentionality in terms of two basic properties: the first is referred to as "direction of fit" and it is specified as world-to-mind. What Searle means is that for a certain intention to be successful, conditions *in the world* must conform to the conditions specified by the intentional state *in the mind*. The second property is referred to as "direction of causation" and it is this time specified as mind-to-world. By that Searle is mainly expressing the fact that it is the intentional state *in the mind* that causes the movement of the agent *in the world*.

We should note, that despite their differences, for Searle both "prior intention" and "intention in action" are essentially representational phenomena. In both cases the intention – as an internal representational state – causes the agent's movement – as an external physical state in the world. The difference is that in the case of "intention in action", the internal intentional state and the external movement become indistinguishable. To highlight this difference, Searle suggests that "intention in action" presents rather than represents its relevant conditions of satisfaction. But this change of terminology from "representation" to "presentation" does not seem to imply much in essence. Presentations are simply "a special subclass of representations" (ibid. 46).

So coming back to Searle's account of intentional activity – that, as we show, he conceptualises as a mind-world transaction – the relationship between the "prior intention" and the "intention in action" can be described by the diagram: Fig. 2.5.

My suggestion is the following: Accepting that agency is about causal events in the physical world rather than about representational events in our mental world, it follows that if an association between agency and intentionality can be made, it has to be with the type of intentionality here called "intention-in-action". In the case of "prior intention", no such correlation can be made before this intention becomes realised in the world – that is, before it meets its relevant condition of satisfaction. This I argue is because as long as "prior intention" is simply an internal representational state, it has no pragmatic effect in the world. As I will discuss in more detail below, pragmatic effect and as such agency is not a matter of private thought and imagination but of actual practice and being-in-the-world. However, once a "prior intention" is realised in the world and as such acquires pragmatic effects, it is immediately transformed to "intention in action". One may suggest here that in this case the "prior intention" can be seen as the cause of the "intention in action" but this is not necessarily the case. This I argue for the following reasons: Firstly, because in most cases "intention in action" is not preceded by a "prior intention". As Searle observes, "[a]ll intentional actions have intentions in action but not all intentional actions have prior intentions" (1983, 85). Secondly, because even when such a "prior intention" exists, it does not necessarily cause or determine the nature and form of a particular activity. For example, an agent may act differently or even in a manner contradictory to his prior intentions or simply fail to meet in action the conditions of satisfaction necessary for such an intentional state to be realised. Finally, even when a prior intention is successfully realised and as such can be argued as

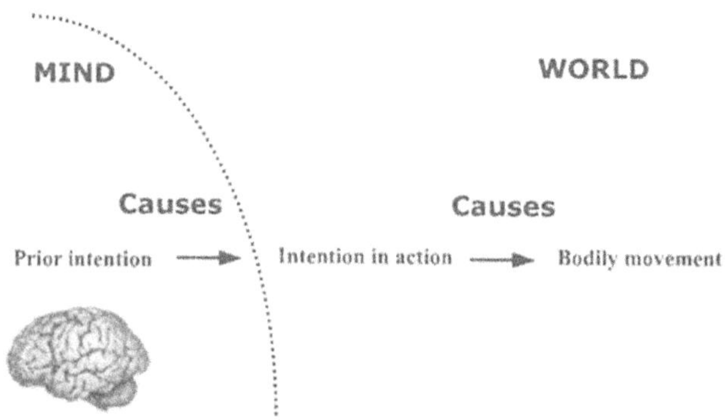

Fig. 2.5 Causal relationship between 'prior intention' and 'intention in action' according to Searle

causing the intention in action, it is already itself being shaped by what I will discuss in a moment as the 'Background'.

As such, I want to suggest that the observed association between agency and intentionality makes proper sense only if conceived as being an association between agency and "intention in action". This minor shift in perspective has some important implications for the meaning of agency because in this case, intention no longer comes before action but it is *in the action*. The activity and the intentional state are now inseparable. As I intend to show in the following, in this case the boundary between the mental and the physical collapses. That means that "intention in action" is not an internal property but a component of extended cognition. Consequently, it constitutes and is constituted both by persons and things and as such it cannot be used as a criterion for ascribing agency to the human component of material engagement. To explicate this claim, I will now turn to discuss 'the Background'.

Searle defines the 'Background' as "a set of non-representational mental capacities that enable all representing to take place" (1983, 143). The Background is the reason that intentional states have the conditions of satisfaction that they do and are the states that they are. To illustrate this, think of what is necessary in order for the potter to form the intention to shape a pot. Think about the number of biological and cultural resources, that he or she must bring to bear on this task, simply to form the intention to perform this task. But without these resources the potter could not form the intention at all.

I have argued previously that in order to understand agency we need to understand action and in order to understand action we need to understand its causal antecedents and in order to do that, we need a *chrono-architecture* of action. I suggest that in this attempted temporal stratigraphy of action the Background offers an artificial, yet much needed, functional boundary. More simply, the clay and its physical affordances, as the basic element of the potter's Background, stands before and causes the potter's intention *about* growing a vessel out of it. And to grow a vessel may take a number of activations or 'presentations' but very few, if any, internal representations – in the computational sense.

Indeed as one may also suggest from a developmental perspective, engagement always precedes intentionality. The child will open the door and learn its capabilities before formulating an intention about the opening of the door. In this sense, the opening of a door is not in itself an intentional state but a part of what Searle calls "local Background" and distinguishes from what he calls the "deep Background". However, what the notion of Background precisely implies in terms of the mind-brain-world connection remains unclear. Searle here, being trapped in an essentially internalist–representationalist view of human mind and intentionality, often appears to be puzzled about where exactly to draw the boundary of human cognition in respect to the 'Background' and how exactly to conceptualise the nature of its properties. He finally settles the issue by calling it 'preintentional' meaning, something that is neither truly mental nor physical.

It comprises the various kinds of 'know-how' – rather than of 'knowing that' – against which intentional states arise:

> The Background, therefore, is not a set of things nor a set of mysterious relations between ourselves and things, rather it is simply a set of skills, stances, pre-intentional assumptions and presuppositions, practices and habits. And all of these, as far as we know, are realised in human brains and bodies (1983, 154).

But where should we look in order to discover this know-how of pottery making? Where, for example, is the knowledge about the precise amount of clay needed to construct the intended pot? Where are the visual categories and morphological prototypes that motivate the shape and form of the vessel?

I am asking *where* rather than *what* because, having previously discussed the temporal anatomy of action, it is now the topology or spatial localisation of agency in particular and of mind in general, that needs further thinking. The thing to note is that in terms of cognitive topology – that is, the question of where those cognitive processes reside – no *a priori* hierarchy can be argued between the potter's brain/body/wheel/clay. For example, the cognitive map of knowledge and memory may well be extended and distributed in the neurons of the potter's brain, the muscles of the potter's body, the motion of the sense organs scanning the surrounding environment for relevant information, the affordances of the potter's wheel – enabling or constraining the discovery of that information – the material properties of the clay, the morphological and typological prototypes of existing vessels as well as the general social context in which the activity occurs. The above components can be broken down further but none of them can be argued as determining the contours of activity in isolation.

Even if one adopts a strictly computationalist view reducing the multifaceted experience of pottery making to a linear, albeit complex, problem solving operation, one would still have to confront the question of boundaries. In fact it would not take too long before realising that important parameters of the problem are enacted outside the potter's head and largely below the conscious threshold of the potter's attention. For indeed, the physical resource be that of clay, wheel, water or instrument, is not simply used by the potter's body following the command of the potter's brain. The physical resources are fully integrated into the functioning and movement of the agent. Clearly it is now the system, the phenomenological compound of brain, body and resource that articulates the boundaries of this intelligent problem-solving ensemble (see also Hutchins 1995; Kirsh 1995, 1996). Of course problem solving is a very poor concept to describe the complexity of the act; it leaves too much out that are of real significance. Nonetheless, my point here is simply to show that even a computational perspective cannot stand outside the loop.

I believe that a much better understanding of the 'Background' can be gained if we view the issue of intentionality from the perspective of the Material Engagement approach. Seen from this angle, the 'Background' becomes a part of the mind or what we may call an *extended intentional state. This implies that the objects and material structures that constitute this 'Background' can be argued to project towards me as much as I project towards them*[2]. In other words, the line between human intention and material affordance becomes all the more difficult to draw. In fact we might even suggest that in certain cases, human intentionality identifies with the physical affordance. The mediational potential of a certain artefact in a quite significant way shapes (both in the positive and negative sense of enabling and constraining) the nature of human intentions.

As is the case with the general issue of human cognition, so it appears to be also in the case of intentionality that "some of our deeply felt assumptions about intentionality, at least as a property of individual minds alone, may be mistaken" (Gibbs 2001, 121). As with many other dimensions of the human mind, intentionality should be understood as a distributed, emergent and interactive phenomenon rather than as a subjective mental state. The artefact should not be construed as the passive content or object of human intentionality but as the concrete substantiating instance that brings forth the intentional state. The world of things elicits and actualises intentionality according to the 'situational affordances' (Gibson 1979; Knappett 2004, 2005) of a given context of engagement.

The 'Background' is where intentionality and the extended mind hypothesis collide. That means that as long as the 'Background' is considered as the *sine qua non* of intentionality, the latter cannot be considered as an internal and purely mental property. But if intentionality is not an internal property, it cannot be used as the criterion for the attribution of agency to humans.[3]

5

It is against this conceptual background that the argument *for* material agency is built. The argument is not for an either/or choice between human and material agency nor for extending a human property to the realm of materiality. The argument is that agency is not a property but the emergent product of the 'irreducible tension of mediated activity' (Wertsch 1998). Within this situated dialectic of activity, material or human predications of agency make sense only from the perspective of dynamic spatio-temporal relations. An agent is defined as "any element which bends space around itself, makes other elements dependent upon itself and translates their will into a language of its own" (Callon & Latour 1981, 286). This is a condition that in any given process of material

[2] "The world is inseparable from the subject, but from a subject which is nothing but a project of the world, and the subject is inseparable from the world, but from a world which the subject itself projects" (Merleau-Ponty 1962, 430).

[3] We are engaged in what Searle (1983) himself recognises as "Networks of Intentional states", but with the requirement that those networks should be better perceived as actor-networks and as such irreducible to any of the constitutive elements in isolation.

engagement can be equally satisfied both by persons and things, the only difference being, that in the latter case – that of things – this process can be sealed in a 'black box' and sink below the surface of our conscious horizon.

The shaping of the pot becomes an act of collaboration between the potter and the mass of wet clay rapidly spinning upon the wheel. There is a constant tactile but also clearly visible, dynamic tension in the movement of clay. On the one hand, the centrifugal force imparted to the clay by the movement of the wheel and the hands of the potter; and on the other, the skilful guidance of this force by the potter's fingers, raising or pressing down the clay to the desired form. It is at the potter's fingers that the form and shape of the vessel is perceived as it gradually emerges in the interactive tension between the centrifugal force and the texture of the wet clay.

In any given stage of this dynamic operational sequence, the wheel may subsume the plans of the potter and define the contours of activity or at another point serve as a passive instrument for his or her manufacturing purposes. At one moment, movement is effortless and feels like happening to the potter rather than being done by the potter as if totally absorbed into the microstructure of clay. At another moment, the potter is clearly conscious of moving clay around and shaping it, directing the flow of the clay and struggling to control the act and handle the clay. Another important parameter in this unfolding dance of agency is the modality involved (e.g., touch versus vision). The wheel and the wet clay are not simply enabling but also constraining.

In the dynamic tension that characterises the processes of material engagement, sometimes it is the thing that becomes the extension of the person. At other times, however, it is the person that becomes the extension of the material agent. There are no fixed agentive roles in this game but a constant struggle towards a 'maximum grip' (Merleau-Ponty 1962). Agency as an emergent property cannot be reduced to any of the human – nonhuman components of action. It can only be characterised according to that component that at a given moment has the upper hand in the ongoing phenomenological struggle.

With respect to agency there is nothing to be found outside this tension of mediated activity and this is precisely the area to which we should look for its manifestations – human or material. Agency is a property or possession neither of humans nor of nonhumans. Agency is the relational and emergent product of material engagement. It is not something given but something to become realised. In short, as far as the attribution of agency is concerned, what an entity (wheel, sheep or tree see Law & Mol; Jones & Cloke this volume) *is* in itself does not really matter; what does matter is what it becomes and where it stands in the network of material engagement.

Our human sense of agency, useful as an evolutionary or social strategy as it might be, it is to a large extent an illusion. But it is not an illusion in the sense that Wegner (2002; 2003; 2004) describes. Causal agency is not something that you discover by going deeper inside the brain. On the contrary, causal agency is what we may call a 'surface property': it dwells at the interface between brains, bodies and things. It cannot be too strongly emphasised that neither brains nor things in isolation can do much. The constant errors in our agency judgements are simply the price we have to pay for being skillfully immersed in a physical world and at the same time of being able to experience this world from a subjective first-person perspective. It is the price of being human. Agency is in constant flux, an in-between state that constantly violates and transgresses the physical boundaries of the elements that constitute it. Agency is a temporal and interactively emergent property of activity not an innate and fixed attribute of the human condition. The ultimate cause of action in this chain of micro and macro events is none of the supposed agents, humans or non-humans; it is the flow of activity itself.

References

Bateson, G., 1973. *Steps to an Ecology of Mind.* London: Granada.

Brooks, R. A., 1991. Intelligence without representation. *Artificial Intelligence* 47: 39–59.

Callon, M. and Latour, B., 1981. Unscrewing the Big Leviathan. In *Advances in Social Theory and Methodology: Towards an Integration of Micro and Macro-Sociology,* edited by K. Knorr-Cetina and A. V. Cicouvel, pp. 277–303. Routledge, Boston, MA.

Gazzaniga, M. S., 1998. *The Mind's Past.* University of California Press, Berkeley.

Clark, A., 1997. *Being There: Putting Brain, Body and World Together Again.* The MIT Press, Cambridge, MA.

Clark, A. and Chalmers, D., 1998. The Extended Mind. *Analysis* 58(1): 10–23.

Dennet, D. C., 1987. *The Intentional Stance.* The MIT Press, Cambridge, MA.

Gallagher, S., 2005. *How the Body Shapes the Mind.* Oxford University Press, Oxford.

Gallagher, S., 2000. Philosophical conceptions of the self: Implications for cognitive science. *Trends in Cognitive Sciences* 4(1): 14–21.

Gell, A., 1998. *Art and Agency: An Anthropological Theory.* Oxford University Press, Oxford.

Gibbs, R. W., 2001. Intentions as emergent products of social interactions. In *Intentions and Intentionality: Foundations of Social Cognition,* edited by F. Bertram, L. J. Moses and D. A. Baldwin. The MIT Press, Cambridge, MA.

Gibson, J. J., 1979. *The Ecological Approach to Visual Perception.* Houghton Mifflin, Boston, MA.

Howhy, J. and Frith, C., 2004. Can Neuroscience explain consciousness? *Journal of Consciousness Studies* 11: 180–198.

Hutchins, E., 1995. *Cognition in the Wild.* The MIT Press, Cambridge, MA.

Knappett, C., 2005. *Thinking Through Material Culture: An Interdisciplinary Perspective.* University of Pennsylvania Press, Philadelphia.

Knappett, C., 2004. The affordances of things: A post-Gibsonian perspective on the relationality of mind and matter. In *Rethinking Materiality: The Engagement of Mind with the Material World,* edited by E. DeMarrais, C. Gosden and C. Renfrew, pp. 43–51. McDonald Institute Monographs, Cambridge.

Knappett, C., 2006. Beyond Skin: Layering and Networking in Art and Archaeology. *Cambridge Archaeological Journal* 16(2): 239–251.

Kirsh, D., 1995. The intelligent use of space, *Artificial Intelligence* 73(1–2): 31–68.

Kirsh, D., 1996. Adapting the environment instead of oneself. *Adaptive Behavior* 4(3–4): 415–452.

Libet, B., 1985. Unconscious cerebral initiative and the role of conscious will in voluntary action, *Behavioral and Brain Sciences* 8: 529–566.

Libet, B., 1999. Do we have free will? *Journal of Consciousness Studies* 6: 47–58.

Libet, B., Gleason, C. A., Wright, E. W. and Pearl, D. K., 1983. Time of conscious intention to act in relation to onset of cerebral activity (readiness potential). The unconscious initiation of a freely voluntary act. *Brain* 102: 623–642.

Malafouris, L., 2004. The cognitive basis of material engagement: Where brain, body and culture conflate. In *Rethinking Materiality: The Engagement of Mind with the Material World,* edited by E. DeMarrais, C. Gosden and C. Renfrew, pp. 53–62. McDonald Institute Monographs, Cambridge.

Malafouris, L., in press. Linear B as distributed cognition: Excavating a mind not limited by the skin. In material culture. In *Excavating the Mind: Cross-sections Through Culture, Cognition and Materiality*, edited by M. Jessen, N. Johanssen and H. J. Jensen. Aarhus University Press, Aarhus.

Malafouris, L. and C. Renfrew, editors, in press. *The Cognitive Life of Things: Recasting the Boundaries of the Mind.* McDonald Institute Monographs, Cambridge.

Merleau-Ponty, M., 1962. *Phenomenology of Perception.* Routledge, London.

Schiffer, B. and Skibo, J. M., 1997. The explanation of artifact variability. *American Antiquity* 62: 27–50.

Searle, J. R., 1983. *Intentionality:* An Essay in the Philosophy of Mind. Cambridge University Press, Cambridge.

Tsakiris, M. and Haggard, P., 2005. Experimenting with the acting self. *Cognitive Neuropsychology* 22(3/4): 387–407.

Van der Leeuw, S. E., 1993. Giving the potter a choice: Conceptual aspects of pottery techniques. In *Technological Choices: Transformations in Material Cultures since the Neolithic*, edited by P. Lemonnier, pp. 238–288. Routledge, London.

Van der Leeuw, S. E., 1994. Cognitive aspects of technique. In *The Ancient Mind,* edited by C. Renfrew and E. Zubrow, pp. 135–142. Cambridge University Press, Cambridge.

Van Gelder, T., 1995. What might cognition be, if not computation? *Journal of Philosophy* 91: 345–381.

Walter, W. G., 1953. *The Living Brain.* Duckworth, London.

Wegner, D. M., 2002. *The Illusion of Conscious Will.* The MIT Press, Cambridge, MA.

Wegner, D. M., 2003. The mind's best trick: How we experience conscious will. *Trends in Cognitive Science* 7: 65–69.

Wegner, D. M., 2004. Precis of the illusion of conscious will. *Behavioral and Brain Sciences* 27: 649–692.

Wertsch, J. V., 1998. *Mind as Action.* Oxford University Press, New York, NY.

Wittgenstein, L., 1953. *Philosophical Investigations.* Basil Blackwell, Oxford.

Chapter 3
Material Agency, Skills and History: Distributed Cognition and the Archaeology of Memory

John Sutton

Introduction

If cognition is distributed as well as embodied, then explanation in cognitive science must often highlight more or less transient extended systems spanning embodied brains, social networks or resources and key parts of the natural and the cultural world. These key parts of material culture are not simply cues which trigger the truly cognitive apparatus inside the head but instead form "a continuous part of the machinery itself", as "systemic components the interaction of which brings forth the cognitive process in question" (Malafouris, 2004:58). On this view, cognitive science is thus not just the study of the brain: indeed, even neuroscience cannot be the study of the brain alone, for brains coupled with external resources may have unique functional and dynamical characteristics apparent only when we also attend to the nature of those resources and the peculiarities of the interaction. This chapter argues that if cognition is indeed thus distributed, then cognition is also historical and heterogeneous and must also be analysed diachronically and differentially. If mind is extended, that is to say, then historical cognitive sciences are essential to the interdisciplinary enterprise.

This is not just because individual brains themselves are "biosocial organs permeated by history" (Cowley, 2002:75) but also on the longer scale because of dramatic historical diversity in the nature, properties and use of cognitive artefacts. According to Andy Clark, "the single most important task" for "a science of the bio-technological mind" is to understand "the *range and variety* of types of cognitive scaffolding and the *different* ways in which non-biological scaffolding can augment (or impair) performance" (Clark, 2002:29, my italics). Unique historical and cultural features of human beings extended cognitive make-up are thus not accidental extras added to a basic biologically given mind. Rather, such changing media, objects, routines, institutions and practices have

J. Sutton
Macquarie Centre for Cognitive Science, Macquarie University, Sydney, NSW 2109, Australia
e-mail: jsutton@maccs.mq.edu.au

C. Knappett, L. Malafouris (eds.), *Material Agency*,
DOI: 10.1007/978-0-387-74711-8_3, © Springer Science+Business Media, LLC 2008

long been integral parts of the coordinated, interactive cognitive systems in which our characteristic plasticity is revealed, engaged and transformed.

Although Clark and other enthusiasts of distributed cognition push on to analyse our couplings with *new* cognitive technologies, there is nothing 'post-human' about the framework itself: if we are natural-born cyborgs now, we always have been. So while some historical, anthropological and archaeological investigations of independent interest can be given a new twist in the light of distributed cognition, they should also help us further develop specific ideas within that theoretical framework (Sutton, forthcoming, a). The nature and extent of diversity in activities of remembering and reasoning, imagining and decision-making, acting and feeling has to be tested across detailed case studies of specific historical periods and cultural contexts. Because neural, bodily, material and social resources can *complement* one another while retaining their own dynamics in making their distinct contributions to integrated cognitive systems, the extent of such integration with external resources varies on a range of dimensions: the context-dependence of flexible intelligent activity itself varies with context, and on occasion – for some individuals or in some unique cultural situations – some activity of embodied brains will be relatively shielded from their environment.

A key current task, then, is the identification of these significant dimensions of variation in constructing better typologies of distributed cognitive systems. What are the synchronic and diachronic principles of coordination between diverse components? What are the different forms of coupling, involving distinct forms of availability and use of external resources? How truly interactive are particular emergent systems and how durable? Cross-cultural and historical data, understandably, have not typically been consulted by those theorists currently working towards a multidimensional framework by addressing these questions (Clark and Chalmers, 1998; Kirsh, 2006; Poirier and Chicoisne, 2006; Sutton, 2006; Wilson and Clark, forthcoming). But alongside relevant work in more recent cultural and cognitive history, which I discuss below, cognitive archaeology can contribute directly to this task both by offering detailed case studies and by broadening theoretical horizons in cognitive science. There are rich resources for these debates, resources of which the broader cognitive scientific community should be better aware, in research on memory in traditions of archaeology which have not explicitly or deliberately engaged with distributed cognition (Alcock, 2002; Ingold, 1998:40–42; Olsen, 2003; Rowlands, 1993; Van Dyke and Alcock, 2003; Williams, 2004). But for now, I focus on a series of related challenges which have already been put directly to the distributed cognition/ extended mind frameworks by cognitive archaeologists, challenges which for the purposes of this chapter I take to be crystallised in an important recent discussion by Lambros Malafouris (2004) of "the cognitive basis of material engagement" (compare also Knappett, 2004, 2005). I develop my case hereby seeking to clarify these challenges and to begin to address them. I deal with issues about history and dynamics, about interactivity and material agency and about skills and skill memory.

Exograms, History and the Cognitive Life of Things

Culturally specific technologies and media, according to Merlin Donald's influential scheme, have constituted part of human cognitive architecture since the upper Palaeolithic period. In particular, changes since then in external symbol systems, which consist of arrays of retrievable traces or 'exograms', have dramatically altered the capacity and operation of human memory (Donald, 1991:308–333; 1998a). Identifying certain common features across the diverse history of external representations – body markings, grave decorations, hieroglyphics, maps, musical scores, writing systems, architectural diagrams and so on – Donald focussed our attention on the new cognitive profiles that characterise creatures (and societies of creatures) who can draw on these exograms in addition to neural engrams. Thoughts and memories, for example, become more durable and more easily transmissible and reformattable across media and contexts and are plugged in to vastly larger databases of inherited knowledge (Donald, 1991:314–9). Mark Rowlands built on Donald's work to argue that much of human memory is essentially (not accidentally) environment-involving and primarily consists in our ability to interface with a range of different collective memory networks (Rowlands, 1999:119–147, also drawing on important work by Rubin, 1995).

Because Donald's substantial treatment of extended memory systems did draw on a wealth of historical and cross-cultural evidence, it drew critical engagement and commentary from cognitive archaeologists interested in material agency (Renfrew, 1998, 2003; Thomas, 1998). Malafouris builds on this work in arguing that Donald's scheme is problematically restricted and incomplete. Due to his "preoccupation with 'exographic storage'", firstly, Donald cannot accommodate cases in which artefacts have "a dynamic cognitive biography", and neglects the unique and idiosyncratic socio-technocultural histories which archaeologists must study (2004:56). As a result, Malafouris suggests, Donald fails to allow for the active role of objects in coordinated interaction, his scheme too rigid in its assumption of "a passive external 'long-term' store" (2004:57). Further, Donald retains too much from classical cognitivism in his focus on straightforward, explicit information transmission, and thus his scheme is blind both to the nonsymbolic cognitive roles of artefacts and to the centrality of know-how and embodied skill in the many diverse ways we "think through things, in action" (2004:57–58). For these reasons, Malafouris thinks we need alternative frameworks to do justice to "the causal efficacy of materiality in the enactment and constitution of a cognitive system or operation" (2004:55). Such views of distributed cognition, he implies, unhelpfully treat the cognitive life of things in artificial isolation from their social life (2004:56).[1]

---

[1] Malafouris repeatedly insists that analysis must include simultaneous attention to material, social and cognitive dimensions, and my responses in this essay seek to show that this is possible. Our views contrast with certain other strands within material culture studies in which the social and the cognitive are decentred or excluded: one recent collection advertises that its

In contrast, I will argue that we can and should, for the moment, rest content with the conceptual resources already in place in the conjunction of the general distributed cognition framework with Donald-style treatments of memory. The required modifications and extensions are more likely to emerge in the hard work of applying these frameworks to detailed cognitive- archaeological (or historical or anthropological or media-theoretic) case studies, than through any further theoretical radicalisation.[2] To make this case here I must address each of Malafouris's concerns in turn: but first, I need to supply some further background on the recent history of the extended mind hypothesis.

Despite its obvious interest to social scientists and archaeologists, Donald-style anti-individualism has been harshly criticised in mainstream philosophy of cognitive science. Adams and Aizawa, for example, assuming that exograms would have to be just the same as engrams to count as cognitive, argue that Donald's careful analyses of the many "ways in which the processing of exograms differs from the processing of engrams" in fact demonstrates that external representations are *non*-cognitive and that the mind is not extended after all (2001:58–59). They thus find it bewildering that Donald agrees with them that internal and external resources differ dramatically in representational format and dynamics, yet still sees cognition as distributed.

The skewed dialectic here results from an overemphasis, by critics and enthusiasts alike, on one route to the extended mind hypothesis, that based on 'Parity'. Clark and Chalmers' 'parity principle' (1998:8) states that an artefact is part of a cognitive process if it performs a function which would thus count as cognitive, if done, in the head. Among other problems with a sole reliance on parity to get the mind out of the skull (Menary, 2006; Sutton, forthcoming, a), this principle seems to allow for uncoupled material-cognitive agency, for things thinking or remembering away quietly by themselves. This encourages critics to scoff – "the black tie I wear at the funeral isn't doing my grieving for me" (Harris, 2004:729) – and tends to discourage attention to full and complex cognitive ecologies. This tendency was perhaps not helped by the way I initially introduced the notion of "the cognitive life of things" (Sutton, 2002) in an analogy which was as yet insufficiently attentive to the subtleties of its sources in Arjun Appadurai's edited collection *The Social Life of Things* (Appadurai, 1986; Kopytoff, 1986; see also Appadurai, 2005). Failing then to stress the assembled, interactive, reciprocal, integrated and integrative nature of distributed cognitive systems, I allowed material agency to appear isolated and self-sufficient. But of course invocations

constituent essays "signal the need to decenter the social within social anthropology in order to make room for the material" (Miller, 2005: back cover).

[2] Here I do not discuss one such radicalisation mentioned by Malafouris, the turn to 'enactivism' and to the work of Maturana and Varela. In another critical response to Malafouris, Mike Wheeler (forthcoming) argues that this tradition is in some tension with the extended mind hypothesis. Although I offer a different reading of Malafouris' discussion of skills and embodied know-how below, my response is compatible with Wheeler's and ends on a similar point.

of exograms and of the extended mind do not entail that external resources do the cognitive work on their own, as if the naked, uncoupled thing were any more autonomous than the naked, uncoupled brain.

Parity does capture the central anti-individualism of the Appadurai-Kopytoff framework: cognitive artefacts, like other socially embedded and culturally transmitted things are not merely passive commodities for the use and pleasure of the active mind (even though they can be thus constructed at specific times within their biographies). But for other central features of Appadurai-style analyses we need to move beyond Parity, to construe the relevant relation between inner and outer resources more as complementarity than as parity. Parity, for Clark, is best taken as an informal test, a place-holder for a "more interesting and plausible argument" for the extended mind, which turns on "the way external elements may play a role different from, but complementary to, the inner ones" (Clark, 1998:99). In the key cases, quite 'alien' and 'disparate' inner and outer components cooperate: while the brain does not need to replicate external forms of storage and computation, for example linguistic forms (Clark, 1997:220; 1998, 99), neither (as Donald shows) must artefacts operate in precisely the same way as brains do nor exactly mimic neural processing profiles.

Complementarity, then, explains why it is natural for cognitive archaeologists or media theorists interested in the extended mind actively to investigate differences between inner and outer resources. But it does not yet sufficiently flesh out the diachronic, biographical or historical aspects of the cognitive life of things. Appadurai and especially Kopytoff had underlined the utility of studying "things-in-motion", entangled in complex networks of use which alter through time and across contexts. Indeed social motion is the medium from which the identification of artefacts can proceed by abstraction, for "all things are congealed moments in a longer social trajectory" (Appadurai, 2006:15). In analyses of case studies like the *kula* system of the Western Pacific and the trade in relics in early medieval Europe, these writers urged attention to the *processes* of engagement or detachment by which, over distinct phases of their "cultural biographies", artefacts come to be more or less integral to social practices and identities. There is often intense labour and difficulty involved, both culturally and cognitively (Kopytoff, 1986:64), in making things act and in acting with things or indeed in stopping them acting or disposing of them. Actors actively couple collective or personal goals and projects to the life of things or play their parts in transferring artefacts from one social zone to another as they are singularised or commodified, as their history is either accumulated and highlighted or stripped away and flattened.

Distinct temporal scales are in play, further, when we analyse the cultural biography of specific things, on the one hand or the broader dynamics of classes or types of thing over a longer historical ebb and flow, on the other. Although cognitive archaeology must also deal with both of these "two dimensions of the temporality of things" (Appadurai, 1986:34–36), Malafouris is right to note that the complex framework of this kind of social theory just has not yet been translated to the study of the cognitive dimensions of the life of things. Even Donald's scheme, based as it is on attention to diverse and tangled forms of coordinated complementarity between distinct inner and outer resources, remains in Malafouris' view too general. Using the example of Linear B clay tablets in the Mycenaean context, Malafouris argues that Donald is too focussed on passive exograms storing discrete contents: it thus neglects the diversity of materials which interact in culturally specific ways with socially embedded actors. The role of mnemo-technical artefacts, in particular, "is far more dynamic and dialogical than the one implied by the notion of a passive external 'long-term' store": instead, such artefacts "engage memory according to the interactional properties which they afford to particular actors in particular settings" (2004:56–57). Donald's "concept of storage" offers little help, Malafouris suggests, in understanding the active and diverse roles of objects in coordinated interaction.

Later I will address Malafouris's positive contributions about the centrality of skills and know-how in thinking-through-things. But first I sketch the resources available within distributed cognition to respond to his reasonable concerns, and discuss a parallel debate about distributed cognition and the cognitive life of things in early modern cultural history which may be instructive and of considerable interest to archaeologists. Initially it is worth pointing out that Donald himself already sought to answer similar criticisms directly. Discussing Thomas' (1998) reading of his work as classically computationalist, Donald confirms that his term 'external symbolic storage' was not "meant to exhaust all the functions of external symbols" and that "the 'storage' function of symbols can neither be isolated from their other functions nor from the minds that use them": further, he sees cognitive artefacts in use as "drawn into a maelstrom of shared cognitive activity in any culture" and argues that "their functions in the larger cultural matrix go well beyond mere storage, because they are in dynamic interaction with the entire cognitive-cultural system" (Donald, 1998b:184).

Yet legitimate worries remain, despite these disclaimers. In repeatedly stressing the radically different properties and dynamics of engrams and exograms,[3] Donald remains primarily focussed on a certain class of external symbol systems, those which do retain as their key function the discrete storage of information even when they also play many other roles. Exograms in such systems have no *intrinsic* dynamics or activity, are not intrinsically integrated with other stored information and do no cognitively work in their standing or dispositional form (when not being currently used, manipulated or activated). This stress makes good sense in the context of mainstream philosophy of cognitive science, where we need to move beyond Parity considerations in motivating

[3] Even in the response to critics quoted above, Donald's way of articulating the complementarity between inner and outer resources is to characterise the "biological memory" with which symbols engage as "a creative, constructive, dynamic force" in sharp contrast to artefacts which "are static things" (1998b:184).

distributed cognition. And Donald's historical point about the cultural-cognitive significance of this core difference also holds good: unlike the constantly moving contents of biological working memory, the products of thinking when reformatted exogrammatically could "be frozen in time, held up to scrutiny at some future date, altered, and re-entered into storage, in a repetitive, iterative process of improvement" (Donald, 1991:316). But this does not mean that we should homogenise all external cognitive resources or see that particular kind of storage as essential and inevitable.

A better way to see the stress on complementarity as a route to distributed cognition is as offering a typology or framework in which many quite different relations (and kinds of relations) between inner and outer resources can be understood (Sutton, 2006; forthcoming, a). External resources with different formats, dynamics and functions permit and encourage quite different kinds of interaction and coupling: the extended mind thesis, thus understood, is more an invitation to give detailed attention to such differences in particular contexts and case studies than a rigid new metaphysics of mind. Not all systems of exograms are meant to be permanent or of unlimited capacity or endlessly reformattable; and not all systems which are intended to endure actually do so in historical and cultural practice. Objects and media, both ancient and modern, may actively change in various ways which shape and influence potential coordinated cognitive interactions. And as soon as we acknowledge that other people may (in certain circumstances) form part of our external memory fields, with their own dynamic engrams potentially acting as exograms for us, it becomes clear that passive external words and images in no way exhaust the media in which cognition and remembering are situated and that materiality can have many different kinds of causal efficacy.[4]

In putting complementarity at the heart of distributed cognition, then, we acknowledge that relations between agents and artefacts may be asymmetric and tangled in different ways and thus that such relations are often dynamically reconfigured or renegotiated over time. But this means that a diachronic dimension will be inevitable in many analyses, because examining the cognitive role of things at a particular isolated time alone will precisely omit the *life* of that thing or class of things. Just as we cannot assume that every individual in some shared cultural context will couple and recouple with cognitive artefacts

[4] In Sutton (forthcoming, a) I build on this point and on more recent work by Andy Clark, to outline an extra twist by which cognitive technologies do not have to be external at all, but include a range of internalised representations and symbol systems which we have learned (historically and developmentally) to manage with both idiosyncratic and culturally specified techniques. Language is just one of these inner prostheses: in that essay, I look at the medieval and Renaissance arts of memory as a further case study which problematises Donald's neat dichotomy between fluid engrams and stable exograms even further, and I suggest that the extended mind thesis can thus encourage us to develop 'a deterritorialized cognitive science which deals with the propagation of deformed and reformatted representations, and which dissolves individuals into peculiar loci of coordination and coalescence among multiple structured media'.

in the same ways, so we cannot assume that every such artefact will retain the same affordances across the various phases of its biography (De Léon, 2003; Knappett, 2004, 2005:35–63). Among the many relevant dimensions here are issues about design, control and power. The assignments of artefacts to roles and to users in distinctive complex cognitive economies may involve processes which are highly contested or which involve unintended consequences of other processes, which then constrain subsequent affordances and possibilities of use and interaction.[5] So an interdisciplinary and historical cognitive science should expect to find deep heterogeneity not only in the nature and properties of media and objects which enter into distributed cognitive systems, but also in the available and actual modes of engagement with such objects over time.

Early Modern Material Agency

Before addressing further issues raised by Malafouris about our taxonomies of memory and about action and know-how, I want to digress by examining a parallel debate in early modern cultural history. Simply announcing that a diachronic dimension is essential or entailed by our conceptual framework, does not of course make it at all easy to identify or provide in particular cases. My claim that we are not far away from having adequate conceptual resources within current versions of distributed cognition will stand or fall on how well they motivate and fit with successful case studies across the disciplines. So only interactive dialogue between the theoretical frameworks and specialist studies of specific rich cognitive ecologies will allow us to extract a more mature and comprehensive approach to the socio-culturally embedded cognitive life of things.

Material culture studies and 'thing theory' have influenced recent cultural and literary-historical studies of early modern Europe, just as they have cognitive archaeology. As in archaeology, some of this work has been explicitly linked to the extended mind and distributed cognition approaches, but more has been motivated by independent developments within the field. A further striking parallel is that such studies have also been criticised for failing to incorporate evidence of diversity and change in the role of cognitive technologies over time across groups and cultures. Jonathan Gil Harris, for example, explicitly complains of the "synchronic bias" of the new object-oriented early modern scholarship (Harris, 2000:114). Because "the current wave of object scholarship" has "largely ignored" Appadurai's stress on "the diachronic trajectories of things through time and space", it remains merely antiquarian and sentimentalist, stuck in a "frozen, glittering present" (Harris, 2000:117–8, 123; 2001:480, 485; compare Klein, 2000).

[5] I draw this Heideggerian sense of 'assignment' from Beth Preston's rich account of cognition and tool use (Preston, 1998).

Yet a number of works on material agency and cognitive artefacts in the early modern period explicitly address issues about historical development and the peculiar cognitive biographies of certain objects. Elsewhere I have discussed Evelyn Tribble's detailed studies of the changing techniques, symbol systems and practices imposed laboriously (and with far from uniform success) on the things, media, buildings and congregations of the new Protestant churches of post-Reformation England, in the wake of new cognitive-mnemonic challenges set by the disappearance of Catholicism's rich multimodal engagement with the sacred (Tribble, 2005a; Sutton, forthcoming, b). Like her groundbreaking reinterpretation of the mnemonic objects and practices of the Renaissance theatre (Tribble, 2005b), this work is explicitly inspired by the distributed cognition frameworks developed by Hutchins (1995) in which neither the information nor the sequences of actions to be remembered need "be explicitly represented anywhere" (Clark, 1997:77). But other work arising from material culture studies also takes early modern scholarship into specifically cognitive domains. The work of Peter Stallybrass and his colleagues, for example, should be of considerable interest to cognitive archaeologists, because while it is thoroughly anchored in a Kopytoff-style biographical approach, it takes this and its other theoretical ingredients into a new and cognitive register by applying them to an ambitious narrative about the changing "materials of memory" in early modern England. I sketch two components of Stallybrass' programme in terms which should be equally pertinent to archaeologists and cognitive scientists.

In brilliant techno-historical detective work, Stallybrass, Roger Chartier and others have reconstructed an early modern technology of memory which had been almost entirely forgotten – 'writing-tables' or 'table-books' with erasable leaves which were increasingly common in England from the 1580 s (Stallybrass, Chartier, Mowery and Wolfe, 2004). These are the 'tables', for example, for which Hamlet calls in order to record the ghost's command. Their pages could be wiped clean with sponges, but faint traces of earlier entries remained. Information in these memory artefacts met most of the criteria set by Clark and Chalmers (1998) for genuine extended cognition: they were ubiquitous across English culture over the period of their heyday; they were portable (small enough to fit in a pocket and be carried about); and they were convenient in use, in that the stylus required was much easier than pen and ink. Yet, in sharp contrast to the permanence of exograms in Donald's paradigmatic external symbol systems, the key characteristic of these tables was the erasability of their contents. Comparing our reliance on erasable memory systems such as computers and electronic organisers, Stallybrass and colleagues however argue that these tables 'shaped and were shaped by a structure of memory different from our own' (2004:410). They trace new pedagogical, practical and rhetorical features of table-books around 1600, examining, for example, commonplacing strategies for the regular redeployment of knowledge and publishers' new strategies for selling calendars or almanacs with extra erasable pages. Different sets of comparisons also became available between these erasable table-books and other technologies of memory. Whereas 'tables' still also meant the stone tablets of God's commandments, the capacity of writing-tables to record information while remaining reusable and open to new pressures opened up new interactive relations with the biological forms of memory, for information held in brains as well as on external surfaces is "vulnerable to the material form on which it is inscribed" (2004:416). Stallybrass and colleagues note too that despite the standard practice of patching together plays from existing models and borrowed fragments, this period also saw the beginning of new concerns about information overload and over-reliance on imitation: the true dramatist is one who does not rely on a table-book (2004:413–4).

This hint about the gradual emergence of a newly individualist conception of imagination as independence from external source materials is filled out in extraordinary detail in Stallybrass' work with Ann Rosalind Jones on "Renaissance clothing and the materials of memory" (Stallybrass, 1993; Jones and Stallybrass, 2000). Drawing on the histories of fashion and of the fetish, Jones and Stallybrass argue that early modern England was a "livery society" in which clothes were "forms of memory that were transmitted". Where we see the person as prior to the clothes worn, then wearers' identities were partly constituted (and constantly renegotiated) by the 'material memories' they wore. Cloth was not only a valuable medium of exchange, but also a key means of incorporation or of binding into social and psychological networks. "The particular power of cloth to effect these networks is closely associated with two almost contradictory aspects of its materiality: its ability to be permeated and transformed by maker and wearer alike; its ability to endure over time. Cloth thus tends to be powerfully associated with memory. Or, to put it more strongly, cloth *is* a kind of memory" (Stallybrass, 1993).

As well as tracing the trajectories – both the paths of social circulation and the uncomfortable or triumphal historical journeys – through their period of specific clothes, textiles, portraits and technologies of fabric, Jones and Stallybrass develop an ambitious diachronic account of "the end of livery". As colonialist contact with new worlds brought abundant exotic goods back to Europe, civilised autonomy gradually became newly imagined as "the detachment of the European subject from those goods". Clothes, along with other increasingly disavowed things, were no longer to be the materialisations of history, memory and desire, for a true individual was "unhampered by fixation upon objects" (2000:11, 269–277). However imperfectly, clothes were reimagined and thus reinhabited as commodities alone.

Questions about how best to theorise and live with memory as literally extended into objects are directly addressed by these early modern case studies. Here, the social and the cognitive life of things are appropriately interwoven; not all cognitive artefacts are passive and not all leave discrete or explicit contents stored in unchanging format; many are not in the information-transmission business in any obvious way at all; the analyses of clothes, artefacts and memory are neatly integrated with approaches to embodiment through the related tradition of "historical phenomenology" which examines the surprising interweaving of humoral psychophysiology with lived early modern bodily experience (Paster, 1993, 2004; Yates, 2006; Sutton, 2007a); and there is room for careful tracing of the changing kinds and levels of material agency among particular kinds of object over time. Despite his critique of other early modern object-oriented history for its blindness to diachronic dimensions, Harris singles out Stallybrass' examination of the transmigrations of costumes between the institutions of church and theatre as a clear example of alert historical method (Harris, 2001:488).

Of course, the artefacts available to cognitive archaeologists are less abundant and often more enigmatic than the diverse things, texts and traces surviving from early modern culture. But, I submit, similar difficulties, nonetheless, arise in each context in seeking to fuse high theory in the philosophy of cognitive science with specialist case studies.[6] Archaeologists may better spy both the potential and the pitfalls of such ambitious attempts in their own field if they keep an eye on their disciplinary neighbours' quests. If I have not provided sufficient detail in this brief account to convince doubters that diachronically sensitive treatments of material agency are possible, I hope at least to have encouraged occasional curious straying. But now I want to return to address some further suggestions made by Malafouris about how we might understand "the constitutive intertwining of cognition with material culture" (2004:57–60). Two outstanding issues remain: the classification of forms of memory and the significance of practical skill.

Skill Memory

Malafouris argues that, as well as shifting the basic analytic unit for theories of memory beyond the individual, cognitive sciences which do justice to material agency will require "a more subtle classification of mnemonic

[6] A distinct but equally rich parallel debate can be found in eighteenth-century studies, where (in the wake of earlier historical work on the vastly expanded cultures of consumption and commercialisation) scholars have for some time addressed the quite different ways in which "material culture formed identity through the ownership and display of luxurious possessions" (Benedict, 2007:193). For more on Enlightenment 'it-narratives' in relation to personal ads and wigs, for example, see respectively Lamb (2004) and Festa (2005). Benedict's analysis of eighteenth-century "thing-poems" is particularly suggestive in demonstrating the pervasiveness of the work of purification, detaching subjects from objects and depsychologising things, at least in an elite culture in which the object "holds no memory, no allegiance, no partiality" (2007:202). A fuller treatment of the history of things from the perspective of the extended mind would seek better to trace historical links between these discussions of early modern and Enlightenment artefacts: the incomplete modern erasure of the mediating work of things counterbalances and drives the invention and maintenance of the ideal autonomous agent understood as a "distinct inner locus of final choice and control" (Clark, 2003; Latour, 1993; Schneewind, 1997:3–11).

operations enacted in the context of material engagement" (2004:57). I interpret this as requiring three shifts: away from a unitary conception of memory; away from thinking of encoding, storage and retrieval as three neatly distinct phases; and away from an exclusive focus on the most explicit, representational or symbolic forms of remembering.[7] Neither Donald nor mainstream cognitive psychologists of memory see memory as a unitary phenomenon, and there is widespread consensus about the constructive nature of remembering, such that there are no sharp lines between encoding, storage and retrieval in terms of the dynamics of information-processing (Sutton, 2003). So far so good, then. Malafouris does not explicitly address the more pervasive distinction between episodic or personal memory and semantic or factual memory, as the two key forms of declarative memory, so I will not discuss the implications of the extended mind hypothesis for these classifications here.[8] Instead I focus on Malafouris' third and main aim, the theoretical decentring of *all* forms of declarative memory in favour of skill memory and know-how. No matter what mental representations we may also construct and employ, he argues, "the efficacy of material culture in the cognitive system lies primarily in the fact that it makes it possible for the mind to operate without having to do so: i.e., *to think through things, in action, without the need of mental representation*" (2004:58, italics in original).

I agree fully with Malafouris that the study of practical skills should be central to the cognitive sciences and to theories of memory and that the extended mind hypothesis should offer us a particularly relevant angle on skills, habits and know-how. Again, I do not see that this emphasis requires us to decentre or to drop our attention to content or to internal and external representations: indeed, as I will argue, successful fusions of phenomenological and cognitive scientific accounts of complex embodied skills will often need precisely to retain the invocation of inner and outer representations, once these are appropriately reconceived as dynamical, active and context-sensitive. So I close

[7] It is not obvious which targets Malafouris has in mind here and I am not certain of this interpretation. Malafouris argues that we should not adopt "ready-made psychological models and classifications" derived from a (classical cognitivist) paradigm in which material culture is treated as "external and epiphenomenal to the mnemonic system proper" (2004:57). But he later suggests that the extended mind hypothesis "qualifies material culture as an analytic object for cognitive science, warranting the use of methods and experimental procedures once applied to internal mental phenomena for use upon those that are external and beyond the skin" (2004:60). I will understand his emphasis as being on the earlier, more revisionary take on existing classifications.

[8] On this point, intriguing and controversial suggestions in Donald's and especially in Rowlands' work (1999:123–9), about the increasingly vestigial role of episodic memory in a world full of exograms, might be countered by a social ontology of memory in which genuinely plural episodic or quasi-episodic, memories are held by groups or by plural subjects, rather than by collocations of individuals (Sutton, forthcoming, c, applying Gilbert, 2004 to the case of memory; compare Wilson, 2005; Tollefsen, 2006).

this essay by identifying one concern about pushing too far the idea that skills and grooved habits are entirely independent of explicit memory, so far that we might end up inadvertently reinstating "the methodological separation between reason and embodiment" which we sought to reject (Malafouris, 2004:59) and mentioning a couple of lines of thought which suggest positive ways to avoid this reinstated dualism.

Malafouris offers a rich descriptive picture of embodied "cognition enacted at the potter's wheel" to illustrate his account of a typical "dance of agency" between technical objects, cultural norms, raw materials, muscles and nerves (2004:59–60). Because this kind of expertise relies on an immense reservoir of practical skill memory, embodied somehow in the fibres and in the sedimented ability to sequence technical gestures appropriately, verbal descriptions of it (by either actors or observers) will be inadequate. As with other complex acquired skills such as those involved in many forms of music, dance and sport, what the expert remembers is in large part consciously inaccessible as well as linguistically inarticulable. One challenging consequences for archaeology is that cognitive ethnography must thus be based in apprenticeship (2004:59); another that, as practitioners of music, dance and sport know from bitter experience, thinking too much about the skills in question can disrupt the flow of successful performance.

These considerations are reminiscent of Hubert Dreyfus' phenomenology of everyday expertise: Dreyfus argues that there are two entirely "distinct kinds of intentional behaviour: deliberative, planned action, and spontaneous, transparent coping" (2002:417). The slow transition from novice to expert status in a practical skill, according to Dreyfus, involves gradually relinquishing any reliance on explicit rules or on conscious deliberation. Like equally extreme interpretations of Merleau-Ponty's phenomenology of embodiment or of dynamical models in cognitive science (his two most relevant theoretical neighbours), Dreyfus thus sees explicit thinking and memory as wholly epiphenomenal in the exercise of genuine expertise: psychological principles or maxims are like training wheels which an expert cyclist has long abandoned.

But this sharp separation of knowing and doing cannot be the whole story about the grooved embodied engagement in material culture of potter or musician or sportsperson. What is striking about the exercise of kinaesthetic memory in such complex acquired skills is that there is never a simple repetition, but rather a contextually appropriate distinctive felt movement dynamics "that is at once both familiar and yet quintessentially tailored kinetically to the particular situation at hand" (Sheets-Johnstone, 2003:71; compare Bartlett, 1932:201–2). The minutely adaptable exercise of embodied skills precisely requires an openness to and awareness of the specifics of a situation. So experts have not in fact entirely isolated and insulated action from thought, but instead have forged active and flexible links between doing and knowing (Stevens and McKechnie, 2005; Sutton, 2005, Sutton, 2007b). To think, with Dreyfus, that all declarative thinking – such as fleeting yet explicit memories of particular relevant past experiences, or a swift, affectively charged reconsideration of the aim of the current practical activity – is merely confabulatory and disconnected from the true embodied sources of that activity, is to risk again cutting embodiment off from reason and making it entirely mysterious that we ever do, imperfectly and fallibly, influence ourselves (compare Selinger and Crease, 2003).

Malafouris is clearly aware of this risk and keen to avoid it. In attacking the "analytically minded archaeologist" who retorts that know-how "should be clearly differentiated from the discursive level of rational thinking", Malafouris asks merely that we do not seek genuine mind *behind* tools and activities, but allows that "mental models, schemata and internal planning procedures" can be "active in the course of any creative process" as long as we "recognize them as the temporally emergent and dynamic products of situated activity" (2004:60). This seems exactly right: thought is not an inner realm behind practical skill, but itself an intrinsic and worldly aspect of real-time engagement with the tricky material and social world. Yet along with other excellent work on habit and skill in the social sciences and philosophy (Casey, 2000; Connerton, 1989; Warnier, 2001), Malafouris' position remains in one respect potentially in tension with any attempt to cash out this perspective in more detail.

Pointing out that many forms of engagement with objects cannot be formulated linguistically, Malafouris cites with approval Renfrew's discussion of implicit memory, in which Renfrew asks archaeologists to move "beyond the deliberately communicative or mnemonic role of some classes of artefact" (Renfrew, 2004:29). Once we see the cognitive system as an integrated whole spanning brains, bodies and the social and artefactual world, and we become sensitive to the variety of operations in which mnemo-technical artefacts and practices are involved, we are rightly rejecting the inevitability of deliberately or linguistically or "explicitly inscribed" information (Malafouris, 2004:57). Malafouris points out that objects can, for example, "force you to remember, without including the content of what precisely is to be remembered" (2004:57). But it does not follow, as Malafouris is tempted to suggest, that no content is involved at all.

In fact, we often need to analyse the transformations and distortions of identifiable representations across a variety of people and media in order to understand particular cognitive and social processes. Thus reconceiving information-transmission as "a gradual propagation of organised functional properties across a set of malleable media" (Hutchins, 1995:312; compare Latour, 1996:58) can give us a grip on how to analyse know-how without falling into that further dualism between skill memory and declarative memory. It is puzzling that experts in many domains, whether acting singly or in group settings, do talk to themselves and to each other in ways which, on our picture, are not to be understood as giving the body explicit instructions from the mind. Not just beginners, but experts in open ball sports, for example, mutter 'watch the ball', while improvising jazz pianists find themselves saying 'sing while you're playing' or, simply, 'jazz hands' (Land and McLeod, 2000; Sudnow, 2001). While such hints and labels are not the top-down reprogramming of a merely mechanical skill, neither are they entirely epiphenomenal noise. The pianist David Sudnow brilliantly characterises them as 'instructional nudges', embodied maxims which have condensed all kinds of history and a variety of 'caretaking practices' into, for example, the arms and shoulders (2001:127–9). Such nudges, which can take an enormous variety of forms, are not instructions, in that they do not do their embodied-cognitive work by imposing their explicit contents on the body: they are material symbols with temporary but crucial causal roles as "a new fulcrum for the control of action" which may (or may not) work temporarily to dampen or recalibrate certain habits and tendencies (Clark, 2005; 2006:294; Sutton, 2007b).

Such verbal or quasi-verbal instructional nudges, operating in the context of anchored, well-navigated embodied routines, take their place alongside an even broader array of other more or less idiosyncratic material, somatic and social anchors. Ed Hutchins has recently sought to add detailed analyses of bodily interaction in shared gesture and talk to the distributed cognition approach (Hutchins, 2005; forthcoming). The spatial and temporal sequencing of gestures, for example, can play a key role in imagining or communicating the dynamics of a complex situation or set of (real or potential) material objects. The one point I have space to note here from Hutchins' persuasive case studies of bodily motion as the medium of thinking itself, rather than merely the indicator of thought, is that again they require us to think of extended and embodied cognitive systems in representational and computational terms. Hutchins demonstrates, for example, that there is a key dimension of commitment along which environmentally coupled, socially situated and materially anchored gestures can vary, with some representings being more tentative (and perhaps more transient), others more firm or tangible. The material or gestural activity in these case studies seeks to integrate or stabilise shared or shareable conceptual information for further manipulation or interaction. This is far from the kind of classical cognitivism which Malafouris (like Hutchins himself) criticises for reducing distributed problem-solving operations to "an isolated individual mental template that precedes and defines the operational sequence" (Malafouris 2004:60): but it is almost as far from wholly anti-representationist versions of phenomenology, enactivism or dynamicism.

In the study of practical know-how as in connection with other questions of material agency, there can be mutually beneficial interaction between cognitive archaeologists and other researchers working in the extended mind/ distributed cognition framework. This is a particularly useful framework within the philosophy of cognitive science just because, when properly understood, it motivates attention to a diverse range of dimensions on which cognitive artefacts and the extended systems to which they contribute, differ. In this chapter, I have attempted to open up some shared space for more ambitious empirical and theoretical development across the disciplines.

Acknowledgments I am particularly grateful to the editors. Some of these ideas were first presented at the workshop 'Extended Mind 2: just when you thought it was safe to go back in the head' at the University of Hertfordshire in July 2006: my thanks to Richard Menary and all those who made comments and suggestions then. I also warmly acknowledge the contributions of Amanda Barnier, Ed Cooke, Doris McIlwain and Lyn Tribble, my collaborators on various parts of the developing framework outlined in this chapter.

References

Alcock, S., 2002, *Archaeologies of the Greek Past: Landscape, Monuments, and Memories.* Cambridge University Press, Cambridge.

Appadurai, A., 1986, Introduction: commodities and the politics of value. In *The Social Life of Things: Commodities in Cultural Perspective,* edited by A. Appadurai, pp. 3–63. Cambridge University Press, Cambridge.

Appadurai, A., 2005, Materiality in the field of Anthropology. In *Commodification: Things, Agency, and Identities (The Social Life of Things revisited),* edited by W.M.J. van Binsbergen and P.L. Geschiere, pp. 55–62. Münster, Lit Verlag.

Appadurai, A., 2006, The thing itself. *Public Culture* 18:15–21.

Bartlett, F.C., 1932, *Remembering: A Study in Experimental and Social Psychology.* Cambridge University Press, Cambridge.

Benedict, B.M., 2007, Encounters with the object: advertisements, time, and literary discourse in the early eighteenth-century thing-poem. *Eighteenth-Century Studies* 40: 193–207.

Casey, E. 2000, The ghost of embodiment: on bodily habitudes and schemata. In *Body and Flesh,* edited by D. Welton, pp. 207–225. Blackwell, Oxford.

Clark, A. 1997, *Being There: Putting Brain, Body, and World Together Again.* MIT Press, Cambridge, MA.

Clark, A. 1998, Author's response: review symposium on *being there. Metascience* 7:95–103.

Clark, A., 2002, Towards a science of the Bio-Technological mind. *International Journal of Cognition and Technology* 1:21–33.

Clark, A., 2003, *Natural-Born Cyborgs: Minds, Technologies, and the Future of Human Intelligence.* Oxford University Press, Oxford.

Clark, A., 2005, Word, Niche, and Super-Niche: how language makes minds matter more. *Theoria* 20:255–268.

Clark, A., 2006, Material symbols. *Philosophical Psychology* 19:291–307.

Clark, A., and Chalmers, D., 1998, The extended mind. *Analysis* 58:7–19.

Connerton, P., 1989, *How Societies Remember.* Cambridge University Press, Cambridge.

Cowley, S.J., 2002, Why brains matter: an integrational perspective on *the symbolic species. Language Sciences* 24:73–95.

De Léon, D., 2003, The cognitive biographies of things. *Lund University Cognitive Studies* 103.

Donald, M., 1991, *Origins of the Modern Mind: Three Stages in the Evolution of Culture and Cognition.* Harvard University Press, Cambridge, MA.

Donald, M., 1998a, Hominid enculturation and cognitive evolution. In *Cognition and Material Culture: The Archaeology of Symbolic Storage,* edited by C. Renfrew and C. Scarre, pp. 7–17. McDonald Institute for Archaeological Research, Cambridge, UK.

Donald, M., 1998b, Material culture and cognition: concluding thoughts. In *Cognition and Material Culture: The Archaeology of Symbolic Storage,* edited by C. Renfrew and C. Scarre, pp. 181–7. McDonald Institute for Archaeological Research, Cambridge, UK.

Dreyfus, H., 2002, Intelligence without representation: the relevance of phenomenology to scientific explanation. *Phenomenology and the Cognitive Sciences* 1:367–383.

Festa, L., 2005, Personal effects: wigs and possessive individualism in the long eighteenth century. *Eighteenth-Century Life* 29:47–90.

Gilbert, M. 2004, Collective epistemology. *Episteme* 1:95–107.

Harris, J.G., 2000, The new new historicism's *Wunderkammer* of objects. *European Journal of English Studies* 4:111–123.

Harris, J.G., 2001, Shakespeare's hair: staging the object of material culture. *Shakespeare Quarterly* 52:479–491.

Harris, R., 2004, Integrationism, language, mind, and world. *Language Sciences* 26:727–739.

Hutchins, E., 1995, *Cognition in the Wild.* MIT Press, Cambridge, MA.

Hutchins, E., 2005, Material anchors for conceptual blends. *Journal of Pragmatics* 37:1555–1577.

Hutchins, E., 2006, Imagining the cognitive life of things. Presented at workshop, *The cognitive life of things: recasting the boundaries of the mind.* McDonald Institute for Archaeological Research, Cambridge. Online at: http://liris.cnrs.fr/enaction/docs/documents2006/ImaginingCogLifeThings.pdf

Ingold, T., 1998, From complementarity to obviation: on dissolving the boundaries between social and biological anthropology, archaeology and psychology. *Zeitschrift für Ethnologie* 123:21–52.

Jones, A.R. and Stallybrass, P., 2000, *Renaissance Clothing and the Materials of Memory.* Cambridge University Press, Cambridge, UK.

Kirsh, D., 2006. Distributed cognition: a methodological note. *Pragmatics and Cognition* 14:249–262.

Klein, K.L., 2000, On the emergence of *memory* in historical discourse. *Representations* 69:127–150.

Knappett, C., 2004, The affordances of things: a post-Gibsonian perspective on the relationality of mind and matter. In *Rethinking Materiality: The Engagement of Mind with the Material World,* edited by E. DeMarrais, C. Gosden, and C. Renfrew, pp. 43–51. McDonald Institute for Archaeological Research, Cambridge.

Knappett, C, 2005, *Thinking Through Material Culture: An Interdisciplinary Perspective.* University of Pennsylvania Press, Philadelphia.

Kopytoff, I., 1986, The cultural biography of things: commoditization as process. In *The Social Life of Things: Commodities in Cultural Perspective,* edited by A. Appadurai, pp. 64–91. Cambridge University Press, Cambridge, UK.

Lamb, J., 2004, The Crying of Lost Things. *ELH* 71:949–967.

Land, M.F. and McLeod, P., 2000. From eye movements to actions: how batsmen hit the ball. *Nature Neuroscience* 3:1340–5.

Latour, B., 1993, *We Have Never Been Modern.* Harvard University Press, Cambridge, MA.

Latour, B., 1996, Cogito Ergo Sumus! Or, psychology swept inside out by the fresh air of the upper deck: review of Hutchins 1995. *Mind, Culture, and Activity* 3:54–63.

Malafouris, L., 2004, The cognitive basis of material engagement: where brain, body and culture conflate. In *Rethinking Materiality: The Engagement of Mind with the Material World,* edited by E. DeMarrais, C. Gosden, and C. Renfrew, pp. 53–62. McDonald Institute for Archaeological Research, Cambridge, UK.

Menary, R., 2006, Attacking the bounds of cognition. *Philosophical Psychology* 19:329–344.

Miller, D. (ed.), 2005, *Materiality.* Duke University Press, Durham NC.

Olsen, B., 2003, Material culture after text: re-membering things. *Norwegian Archaeological Review* 36:87–104

Paster, G.K., 1993, *The Body Embarrassed: Drama and the Disciplines of Shame in Early Modern England.*Cornell University Press, Ithaca.

Paster, G.K., 2004, *Humoring the body: emotions and the Shakespearean stage.* Chicago University Press, Chicago.

Poirier, P., and Chicoisne, G., 2006, A framework for thinking about distributed cognition. *Pragmatics and Cognition* 14:215–234.

Preston, B., 1998, Cognition and tool use. *Mind & Language* 13:513–547.

Renfrew, C., 1998, Mind and matter: cognitive archaeology and external symbolic storage. In *Cognition and Material Culture: The Archaeology of Symbolic Storage*, edited by C. Renfrew and C. Scarre, pp. 1–6. McDonald Institute for Archaeological Research, Cambridge.

Renfrew, C., 2003, *Figuring It Out: The Parallel Visions of Artists and Archaeologists*. London, Thames and Hudson.

Renfrew, C., 2004, Towards a theory of material engagement. In *Rethinking Materiality: The Engagement of Mind with the Material World*, edited by E. DeMarrais, C. Gosden, and C. Renfrew, pp. 23–31. McDonald Institute for Archaeological Research, Cambridge.

Rowlands, Mark, 1999, *The Body in Mind: Understanding Cognitive Processes*. Cambridge University Press, Cambridge.

Rowlands, Michael, 1993, The role of memory in the transmission of culture. *World Archaeology* 25:141–151.

Rubin, D.C., 1995, *Memory in Oral Traditions: The Cognitive Psychology of Epic, Ballads, and Counting-out Rhymes*. Oxford University Press, Oxford.

Schneewind, J.B., 1997, *The Invention of Autonomy*. Cambridge University Press, Cambridge.

Selinger, E.M. and Crease, R.P., 2003, Dreyfus on expertise: the limits of phenomenological analysis. *Continental Philosophy Review* 35:245–279.

Sheets-Johnstone, M., 2003, Kinesthetic memory. *Theoria et Historia Scientiarum* 7:69–92.

Stallybrass, P., 1993, Worn worlds: clothes, mourning, and the life of things. *Yale Review* 81:183–207.

Stallybrass, P., Chartier, R., Mowery, J.F. and Wolfe, H., 2004, Hamlet's tables and the technologies of writing in renaissance England. *Shakespeare Quarterly* 55:379–419.

Stevens, C. and McKechnie, S., 2005, Thinking in action: thought made visible in contemporary dance. *Cognitive Processing* 6:243–252.

Sudnow, D., 2001, *Ways of the Hand*. 2nd edition, MIT Press, Cambridge, MA.

Sutton, J., 2002, Porous memory and the cognitive life of things. In *Prefiguring Cyberculture: An Intellectual History*, edited by D. Tofts, A. Jonson, and A. Cavallero, pp. 130–141. Power Publications and MIT Press, Sydney and Cambridge, MA.

Sutton, J., 2003. Memory. *Stanford Encyclopedia of Philosophy*, online at http://plato.stanford.edu/archives/spr2003/entries/memory/.

Sutton, J., 2005. Moving and thinking together in dance. In *Thinking in Four Dimensions: Creativity and Cognition in Contemporary Dance*, edited by Robin Grove, Catherine Stevens, & Shirley McKechnie, pp. 50–56. Melbourne University Press, Melbourne.

Sutton, J., 2006, Distributed cognition: domains and dimensions. *Pragmatics and Cognition* 14:235–247.

Sutton, J., forthcoming, a, Exograms and interdisciplinarity: history, the extended mind, and the civilizing process. In *The Extended Mind*, edited by R. Menary. Ashgate, Aldershot.

Sutton, J., 2007a, Spongy brains and material memories. In *Embodiment and Environment in Early Modern England*, edited by M. Floyd-Wilson and G. Sullivan, pp. 14–34. Palgrave Macmillan, London.

Sutton, J., forthcoming, b, Remembering. In *The Cambridge Handbook of Situated Cognition*, edited by M. Aydede and P. Robbins. Cambridge University Press, Cambridge.

Sutton, J., 2007b, Batting, habit, and memory: the embodied mind and the nature of skill. sport in society 10: 763–786.

Thomas, J., 1998, Some problems with the notion of external symbolic storage, and the case of Neolithic material culture in Britain. In *Cognition and Material Culture: The Archaeology of Symbolic Storage*, edited by C. Renfrew and C. Scarre, pp. 149–156. McDonald Institute for Archaeological Research, Cambridge.

Tollefsen, D.P., 2006, From extended mind to collective mind. *Cognitive Systems Research* 7:140–150.

Tribble, E., 2005a, The chain of memory: distributed cognition in early modern England. *Scan: Journal of Media Arts Culture* 2, online at http://scan.net.au/scan/journal/display.php?journal_id = 53.

Tribble, E., 2005b, Distributing cognition in the globe. *Shakespeare Quarterly* 56:135–155.

Van Dyke, R.M. and Alcock, S. (eds.), 2003, *Archaeologies of Memory*. Blackwell, Oxford.

Warnier, J-P., 2001, A Praxeological approach to subjectivation in a material world. *Journal of Material Culture* 6:5–24.

Wheeler, M., forthcoming, Minds, things, and materiality. In *The Cognitive Life of Things*, edited by C. Renfrew and L. Malafouris. McDonald Institute for Archaeological Research, Cambridge.

Williams, H. (ed.), 2003, *Archaeologies of Remembrance: Death and Memory in Past Societies*. Kluwer Academic Publishers, Dordrecht.

Wilson, R.A., 2005, Collective memory, group minds, and the extended mind thesis. *Cognitive Processing* 6:227–236.

Wilson, R.A. and Clark, A., forthcoming, How to situate cognition: letting nature take its course. In *The Cambridge Handbook of Situated Cognition*, edited by M. Aydede and P. Robbins. Cambridge University Press, Cambridge, UK.

Yates, J., 2006, What are "Things" saying in renaissance studies? *Literature Compass* 3:992–1010.

Chapter 4
The Actor-Enacted: Cumbrian Sheep in 2001

John Law and Annemarie Mol[1]

Introduction

It is a sheep. You can see the photo. But is it an actor? That is the question. Questions of this kind, questions about agency, are usually asked as part of a search for explanation. What is the origin of an event? It is as if the aim of scholarship was to write whodunits. Hidden in the background is a 'structure' 'agency' divide. Are the determinants of this, that or the other event located in existing social structures or do they lie in original, reflexive human beings?

Obviously a sheep is not a human being. This means that like many of the other contributors to this book, if we ask whether or not it is an actor we are starting to destabilise the structure-agency dualism. For whatever it is, a sheep is not reflexive in the way usually imagined by social science. We also start to erode another common feature of the structure-agency dualism, a distinction between mastery and being-mastered. Does a sheep exert mastery, does it control? Or is it simply being pushed around? The answer, we will see, does not fit this division.

J. Law
Centre for Science Studies and Department of Sociology, Lancaster University, Lancaster LA1, 4YN, UK
e-mail: j.law@lancaster.ac.uk

[1] We would like to thank: Nick Bingham, Steve Hinchliffe, Ingunn Moser, Jeannette Pols and Vicky Singleton for help in thinking about agency, animals, and this paper.

C. Knappett, L. Malafouris (eds.), *Material Agency*,
DOI: 10.1007/978-0-387-74711-8_4,

One of the traditions that allows us to escape structure-agency dualism is that of material semiotics. This disentangles agency from intentionality. Within material semiotics, an entity counts as an actor if it makes a perceptible difference.[2] Active entities are relationally linked with one another in webs. They make a difference to each other: they make each other *be*. Linguistic semiotics teaches that words give each other meaning. Material semiotics extends this insight beyond the linguistic and claims that entities give each other being: that they enact each other. In this way of thinking agency becomes ubiquitous, endlessly extended through webs of materialised relations. But where to localise agency in such a web? Where to pin it down? This becomes a matter of attribution, *post hoc* and after the action.[3] In telling stories about events, some entities are detached from their background and called 'actors'. They are made to conceal and stand for the web of relations that they cover. They become the place where explanation, moral, causal, practical, stops.

In the stories that material semiotics makes possible, an actor does not act alone. It acts in relation to other actors, linked up with them. This means that it is also always being acted upon. Acting and being enacted go together. What is more, an enacted-actor is not in control. To act is not to master, for the results of what is being done are often unexpected. This has been said before in the literatures on material semiotics,[4] but frequently this message fails to travel with those literatures and gets forgotten. Perhaps, then, it deserves to be argued again – and in this chapter we will try do this. We will show that acting may be told as a fluid event and that beyond the structure/agency divide the most interesting questions have little to do with mastery. In order to make a difference, a sheep does not need to be a strategist. Neither do you and I.

So, let us revisit our initial question: is a sheep an actor? In order to tackle this question we will unravel its terms. What, then, is 'an actor'? And, at least as difficult: what is a 'sheep'?

Sheep Enacted

The picture above is not of a sheep-in-general, but a specific sheep. Uncoincidently, it comes from Cumbria and we will assume that we are in the middle of March 2001, March 15th, to be precise.[5] This was a special moment in the life and death of a Cumbrian sheep, for foot and mouth disease had taken hold in the UK and it was particularly virulent in Cumbria. It was government policy to

---

[2] In this sentence we combine tropes from Latour (1988) who talks of actors as entities that act and Haraway (1991), who talks of making a difference.

[3] Callon (1986).

[4] See, for instance, Akrich (1992) and (Law: 2003).

[5] It was difficult to take photographs of Cumbrian sheep on that date, since the countryside was closed to the general public for disease control reasons.

eradicate the disease by slaughter. This meant that animals (sheep, cows and pigs) on 'infected premises' and those premises counted as 'dangerous contacts' were being killed.[6] The reasoning was that if you kill the animals, you kill the virus too. But notwithstanding the slaughter, the disease was still spreading like wildfire. A week earlier 126 premises had been infected across the nation. On the 15th, the figure was 250. Government epidemiologists were privately saying that the epizootic was not under control.[7] And locally it was spreading through the farms and the hills of the Lake District. The government was in a panic. Policy was changing week by week. A three kilometre 'pre-emptive cull' of sheep was in the works,[8] and it was announced on the afternoon of the 15th.[9]

At this particular point, then, Cumbria, March 15th, 2001, the sheep in our picture is not only 'in a picture'. It finds itself at the cross-roads of a diverse set of practices. In each of these practices 'a sheep' *is* something different. Each of these practices *enacts* 'sheep' in a different way. Let us present four versions of a sheep.

The Veterinary Sheep

First, in veterinary practice the sheep is a potential *host* for the foot and mouth virus.

> FMD is probably the most contagious virus known in mammals. ... Cattle, sheep, goats, pigs and buffalo are the most important susceptible species. ... In smaller ruminants, such as sheep ... the disease often takes a mild [..] form in adult animals. In young animals ... the virus can cause an acute myocarditis resulting in sudden death.[10]

But that a sheep is a *potential* host to a virus, does not mean that it is easy to know whether or not a particular sheep is actually infected. The vet and the farmer may examine an infected sheep very carefully and yet see nothing strange.

> Recognition of the disease among stock remains the most important step and depends crucially on disease awareness by the farmer and good-communications between farmer and veterinarian. For FMD it requires examination of the visible mucous membranes of the conjunctiva, nose, mouth, tongue and eyes and the external surface of the body and limbs. Recognition in cattle and pigs is usually relatively easy but is more difficult in sheep, in which infection can be sub-clinical.[11]

---

[6] For general accounts of the early evolution of the policy see Foot and Mouth Disease 2001: Lessons to be Learned Inquiry (2002, pp 76ff). For a timeline see National Audit Office (2002, Appendix 1, pp 105–111).

[7] National Audit Office (2002, 61).

[8] Foot and Mouth Disease 2001: Lessons to be Learned Inquiry (2002, 88).

[9] The Minister muffed his lines by talking of 'animals' rather than 'sheep' to the horror and the anger of dairy and beef farmers. Cumbria Foot and Mouth Disease Inquiry (2002, 34).

[10] The Royal Society (2002, 18).

[11] The Royal Society (2002, 78).

In addition symptoms of foot and mouth are easily confused with those of a range of other diseases: sheep go lame for all sorts of reasons.[12] So the sheep-host of veterinary practice may look like an ordinary sheep.

> FMD in sheep is difficult to diagnose. Farmers and vets can miss the signs. Infected sheep often display mild symptoms, if any, and suffer from other conditions that may be confused with FMD.[13]

Because clinical symptoms are not all that clear, in veterinary practice the gold standard for establishing whether or not a particular sheep is a host to the foot and mouth virus is a laboratory test. Technicians isolate the virus from a sheep's bodily material and grow it in a tissue culture but this is slow (it takes up to four days). An alternative laboratory test (an ELISA test for viral antigens in the sheep's blood) is quicker (only four hours) but less reliable.[14] In any case, sending samples off to a distant laboratory is time-consuming and in March 2001 time was of the essence: foot and mouth was spreading, sick animals needed to be slaughtered and any delay in slaughter favoured the spread of the disease. So when the vets looked at a suspect sheep, they did not wait for lab results. They ordered its slaughter. Thus while it is 'normal veterinary practice' to wait for lab results, in March 2001 the vets diagnosed the disease on the basis of clinical inspection alone. They were not really supposed to do this – official policy required laboratory confirmation of disease – but policy was lagging behind practice. In the heat of the moment, clinical diagnosis had elbowed the laboratory aside, even though sometimes the laboratory later showed that the wrong diagnosis had been made.[15] In veterinary practice, then, a sheep is a potential host for the foot and mouth virus. But there are two ways to decide whether or not a particular sheep is carrying the disease, by using clinical or laboratory means, and they do not necessarily lead to the same conclusion.

The Epidemiological Sheep

In epidemiology the sheep is enacted differently. It does not come alone, as a body to be diagnosed. Instead, geographically located collections of susceptible animals are treated as the inputs and outputs of *calculations*.[16] In 2001 these calculations defined a collectivity (a premises, usually a farm), and then made

[12] This happened close to the beginning of the outbreak. See Department for Environment Food and Rural Affairs (2002, 22).

[13] Foot and Mouth Disease 2001: Lessons to be Learned Inquiry (2002, 49).

[14] Cumbria Foot and Mouth Disease Inquiry (2002, 49, 55). ELISA is the acronym for enzyme-linked immunosorbent assay. The Royal Society (2002, 76).

[15] Formal policy, in the form of 'slaughter on suspicion' fell into line with practice on March 15th. National Audit Office (2002, 61).

[16] For details see Kao (2002), The Royal Society (2002, 66–71), and Taylor (2003).

assumptions about the infective relations between those premises. They extrapolated from existing to future cases by making assumptions about possible policy interventions. The probability of infection for a given premise was determined by distance on the one hand (the closer the more likely the disease would pass) and a series of 'heterogeneities' including animal susceptibility to infection, length of infectious period, numbers of animals, farm size, the arrangement of fields on a farm and meteorology:[17]

> All the models showed that culling farms neighbouring infected premises would reduce spread of infection and control the epidemic. This was based on the observation that, on average, animals on 34% of premises within a radius of 1.5 km of infected premises came down with FMD.[18]

So, in epidemiology any particular sheep was part of a larger collectivity of animals-on-a-premises that came with two probabilities attached: one, the likelihood of being infected in a unit time; and two, the likelihood of infecting other collectivities of animals, again in a unit time. This epidemio-logic had complex and multiple relations with the logic of veterinary practice.

One. Epidemiology *differed from* the clinical and laboratory practices of veterinarians by substituting probabilities of infection for clinical and/or laboratory diagnoses.[19] Two. But then again, it also *depended on* and included these, first to build its predictions and second, to confirm its findings. Three. The 'slaughter on suspicion' policy displaced laboratory logic and made clinical logic dominant on the farm and this was for epidemiological reasons. The lab was simply too slow to stop the epizootic from spreading. One might therefore say that epidemiology *decided* between the two variants of veterinary practice. Four. On March 15th, it was determined that all sheep within 3 km of infected premises would be slaughtered because calculations predicted that they ran a considerable risk of being infected.[20] For these sheep, slaughter not only did without laboratory diagnosis, but also the clinical diagnosis of the vets was made *irrelevant* as well. A purely epidemiological logic ruled.

Further complexities arose because epidemiology itself was not unequivocal. First, the models grew out of patchy and inadequate data produced in a bureaucracy that was more or less overwhelmed by events.[21] Second, they

---

[17] The list of possible heterogeneities is endless.

[18] Foot and Mouth Disease 2001: Lessons to be Learned Inquiry (2002, 96).

[19] The complex intersections of practices is explored in Mol (2002).

[20] 'The justification for culling contiguous premises was founded on a statistical concept. All the models showed that culling farms neighbouring infected premises would reduce spread of infection and control the epidemic. This was based on the observation that, on average, animals on 34% of premises within a radius of 1.5 km of infected premises came down with FMD. Although culling contiguous premises was a blunt policy instrument, it had the benefit of speed in decision making. It did not depend on the epidemiological groundwork to identify dangerous contacts, which was resource intensive and time consuming.' Foot and Mouth Disease 2001: Lessons to be Learned Inquiry (2002, 96).

[21] See, for instance, Shannon (2002, 5).

were produced in more or less inconsistent statistical models located in different computers.[22] Third, they were in part the product of a more or less ungentlemanly struggle taking place in the rooms of Whitehall between the proponents of the different models.[23] These differences did not have direct bearing on the fate of our Cumbrian sheep[24] but the struggle worked to increase the (already considerable) political temperature.

The Economic Sheep

There were economic dangers too. To set the stage, it is likely that the UK lost around £130m net in meat and livestock exports as a result of the epizootic.[25] What about our sheep? There are a lot of them in the UK: in 1999 around 19 million.[26]

> We having been the major exporters to other EU countries – we are the main supplier of lamb to other European countries – as soon as that volume of 100,000 tonnes or thereabouts is not available, they cannot immediately source additional supplies.[27]

These sheep come onto the market because farmers buy and sell individual sheep quite unsentimentally – this is a part of trying to make a living. But however large the overall numbers, many farming incomes are modest.[28] Farming in the Cumbrian uplands was (and is) marginal at best and in March 2001 farm incomes were much lower than they had been a few years earlier.[29] It was a

---

[22] There were one or two other models as well: here we are simplifying.

[23] The first was a quickly-calculated, relatively simple deterministic, pseudo-geographical model, developed at Imperial College, London. The other was a much more complex, GIS-based, stochastic model with many more heterogeneities that was being run at the government in-house Veterinary Laboratories Agency (VLA). In the middle of March these agreed that the epizootic was growing. Otherwise, they were very different. The VLA model predicted a total of 1000–2000 infected premises by the end of the outbreak. The epizootic would, it said, stop in due course. Much more alarmingly, the Imperial College model predicted 1000 new infective premises each day by mid May. See National Audit Office (2002, 61) and Foot and Mouth Disease 2001: Lessons to be Learned Inquiry (2002, 88).

[24] The 3 km precautionary cull was supported by the proponents of both models.

[25] During this period all exports of meat and live animals were stopped. Foot and Mouth Disease 2001: Lessons to be Learned Inquiry (2002, 133). This was only a small proportion of the cost to the overall economy.

[26] The Royal Society (2002, 12).

[27] House of Commons Select Committee on Agriculture (2001, Answer to Question 499).

[28] House of Commons (2001, Column 94 WH).

[29] Policy Commission on the Future of Farming and Food (2002, 13). See also the comments by Peter Atkinson MP 'Before the foot and mouth epidemic, farming incomes, particularly those of hill farmers, were historically low. Some estimated their annual incomes just before the epidemic at about £4,000, although others would describe that as extremely optimistic. Since the outbreak, incomes have fallen even further..'. House of Commons (2001. Column 90 WH). See, in the same debate, the comments by Alan Beith MP: 'Hill farmers cannot

bad moment for limiting the possibilities for making a profit, yet this was what happened. For buying and selling sheep, even healthy sheep, was constrained by restrictions on movement (no movement at all with out a licence). It was also affected by the fear of UK lowland farmers of introducing foot and mouth to their own farms and by the different tastes of consumers in different parts of the EU.[30] Market prices fell,[31] and for many farmers, coming after a number of years of economic (and often personal) depression, this led to acute economic stress.[32]

To compensate for the poor financial returns for farmers (and also to increase overall EU agricultural production) there were EU CAP (Common Agricultural Policy) support payments.

> In 2000 subsidies accounted for 50%for total output in hill flocks, 42% in upland flocks and 27% in lowland flocks.[33]

These 'headage' payments, the product of compromise between divergent European interests, thus represented a considerable proportion of the income per sheep.[34] They also led to a large national trade in sheep as farmers ensured that their flocks were up to size in the spring for the CAP subsidy.

And then, crucial to the economic enactment of sheep in March 2001, the Ministry paid compensation for sheep slaughtered. This was an important part of disease control policy (and also supported the farmers):

> In some cases, it is likely that the compensation paid to farmers exceeded the amount which they would have expected to obtain for their animals in normal conditions, possibly by substantial amounts. It was judged necessary to pay farmers on a generous basis to ensure their co-operation in the slaughter policy.[35]

So slaughtered sheep were paid for whilst, given market conditions, those not slaughtered represented a considerable financial liability. If the sheep was enacted as an economic entity then slaughter was often a good.

---

continue taking less than the cost of production, which they are doing now and have sometimes had to do in the past.', House of Commons (2001. Column 94 WH) .

[30] Peter Atkinson, MP, in the House of Commons (2001, Column 91 WH).

[31] 'I was talking to a farmer who was selling horned Blackface sheep for carcass export at 270p a kilo before the foot and mouth outbreak; after the outbreak the price is 150p a kilo because there is no export market.' This is Alan Beith, MP. See House of Commons (2001, Column 94 WH).

[32] See the intervention by MP Peter Atkinson at House of Commons (2001, Column 90 WH), and the testimony of Ms Boundy at Mercer (2002, 52).

[33] The Royal Society (2002, 12).

[34] Ashworth, Palmer and Northen (2000, 97). 'Support payments for sheep and cattle under the CAP are paid on a headage basis, meaning that the more animals a farmer keeps, the greater the subsidy he receives. The number of breeding ewes in the uplands increased by around 35% between 1980 and 2000.' English Nature, quoted by the Policy Commission on the Future of Farming and Food (2002, 74).

[35] Foot and Mouth Disease 2001: Lessons to be Learned Inquiry (2002, 132).

The Farming Sheep

But on a farm a sheep is not only an individual animal with an economic value. It is also a member of a flock. This drastically changes any assessment of slaughter. For there is pride in the history of breeding (selecting, caring) that goes into the raising of a flock. Here is a quote from Devon, but the sentiment was widespread in Cumbria too:

> With my family I have been here since 1946, coming with my parents from my grandparents' farm also in the parish. We brought sheep with us as a nucleus for our flock. Although not a pedigree flock, every year the best ewe lambs were selected to replace culled ewes, as had been the practice from grandfather's early days in farming – so bloodlines went back to them.[36]

Flocks involve long-lasting relations of care. The ewes care for the lambs,[37] and the farmers care for the individual sheep and for the flock as something that is more than the sum of its individual members. Note that care is not maudlin: it may be harsh too.[38] But such harshness has its limits. Here is a Cumbrian farmer:

> We are not sentimental about sending our animals to the abattoir when they reach maturity and peak condition but the mass slaughter of young, some very young, animals seems very unnatural and it feels as if we are going to fail in our duty to them.[39]

A *flock* is a collective that cannot or should not be lost. One of the daily practices of farming is taking care of a flock. And that care is inseparable from geography, from topography and from meteorology.

In Cumbria the farmhouse and a few fields often lie at the bottom of the valley but there are also extensive unfenced upland and highland fells for grazing. The sheep live in those uplands and highlands for much of the year. However, since in winter the tops are covered in snow, conditions are harsh and there is little for them to eat, they are moved down in the autumn. Lambs for fattening that will be sold in autumn and the winter are kept on the lower slopes. Older breeding ewes or 'draught sheep' that can no longer live on the high fells are sold to lowland farmers or for slaughter. And then there are the breeding ewes that will lamb in the spring. More breeding ewes may be bought in the late winter[40] but it is the over-wintering breeding ewes and their springtime lambs that form the core of the flock and secure its continuity. These carry it from generation to generation. It is these that live the annual round of tupping,

---

[36] Mercer (2002, 77). This quotation comes from the Devon Inquiry and does not refer to a Lake District flock.

[37] 'One of the worst things was the twin lambs from a sheep that was giving birth as the sheep had been rounded up just minutes earlier. The sheep would have been a good mother as it was really reluctant to leave them and difficult to load.' (Jackson: 2001, 108).

[38] Mol (2006).

[39] Stockdale (2001, 114) .

[40] This is in time for the CAP headage payment.

lambing and shearing. And if they are released in the late spring onto the uplands and the highlands, this also helps to care for the fields around the farm. These, after all, should not be overgrazed and it is good to allow them time to recover during the summer.[41]

But in March 2001, it was illegal to move sheep on or even across a public highway. As a result many were trapped on the lower fells and unable to come to the lowlands. Others were caught in lowland fields grazing on the pasture usually set aside for breeding ewes and their lambs.[42] They could not be sold but there was not enough pasture to go around. This was not only another economic strain on the farmers, but it was also bad for the welfare of the flock.[43]

The foot and mouth epidemic, with the threats and the restrictions it brought, harmed the welfare of every Cumbrian flock. And some flocks, those that were slaughtered to prevent the epizootic from spreading, were permanently destroyed.

> The only thing worse for a farmer than sending his entire healthy flock for slaughter must be to see his flock come down with foot and mouth and to have them shot in front of him and then to be piled up in his farmyard waiting to be burnt or buried.[44]

Farming practices, then, enact sheep as animals tied to time, place, sex and age. No individual sheep matters very much, but as a flock they are of immense and irredeemable value. The worst thing it did, then, this disease, was not to kill individual sheep, but to eradicate entire flocks of them.

Sheep Acting

You cannot learn what a sheep is by staring at a picture. It helps more to unravel the practices in which sheep figure, in which they are enacted in one way or another.[45] If we do this then we do not discover a sheep that is unified and coherent. Instead we find a 'sheep multiple'.[46] This is because a slightly different sheep is done in each practice. Even so, this multiplicity is not a plurality. Instead there are complex and intricate relations between the various versions

---

[41] It cares much less well for the uplands where there was substantial overgrazing, encouraged, according to the critics, by the CAP headage subsidy for breeding ewes. See English Nature, quoted in Policy Commission on the Future of Farming and Food (2002, 74).

[42] They could also be used for cows that were wintering in the byre, but this is just another of the many complexities we leave out.

[43] Hence the importance of the welfare slaughter scheme. For an account of the difficulties faced by hill farmers, see David Curry MP at House of Commons (2001, 96 WH).

[44] Jackson (2001, 109).

[45] There is a small body of work which explores the character of animals and micro-organisms in the context of situated and multiple practices. See Haraway (2003), Bingham (2006) and Hinchliffe (2001). The work of Laurent Thévenot (2001) on moral complexity in practice also touches on relations with animals and indeed sheep.

[46] Mol (2002) talks of the 'body multiple'. For related arguments, see Law (2002).

of a sheep. So the (simplified!) stories of the different versions of the Cumbria sheep in March 2001 both exclude and include each other. The farming sheep was invaluable, outside value, whereas the economic sheep had a price on its head. The farm flock deserved protection, whereas a sheep enacted as an economic entity was more valuable dead than alive. And the epidemiological sheep and the veterinary sheep simultaneously clashed with and depended on one another. At the same time, however varied the 'sheep' they enacted, on a day-to-day basis all these practices somehow came together. Economic accounts and epidemiological calculations appeared on similar kinds of spread sheets and policy took both into account. The vet visited the farm while the farmer, besieged by foot and mouth was isolated and did not move far. Some members of the family went to live in distant holdings to limit the coming and going.[47]

So, if sheep were enacted in different ways, the different 'sheep' also held together. These are sheep *multiple*: more than one but less than many.[48] But if we say that they were 'being enacted' does this mean that they were passive? Here we hit a linguistic obstacle. The English language makes it easy to write sentences that are active or sentences that are passive. But writing somewhere in between 'doing' and 'being done to' is much more difficult. The divide between 'mastery' and 'being mastered' is thoroughly embedded in English and in its neighbouring European languages. Active or passive, control or slavery, the division is an enduring and central Western concern. And it is precisely this way of building the world that here we seek to interfere with. Sheep enacted also act. If sheep are enacted in different versions, this also means that they act in different ways. But how? What is it that a veterinary sheep, an epidemiological sheep, an economic sheep and a farming sheep actually *do*? How does this animal-multiple make a difference? As we answer this question, we will gradually learn more about the second term we are after, the 'actor'.

Sheep and Vets

We have seen that in veterinary practice sheep are enacted as potential hosts for the foot and mouth virus but that it is difficult to know whether a particular sheep is indeed host to the virus or not. Let us look again at one of the quotes we used above to illustrate this:

> FMD in sheep is difficult to diagnose. Farmers and vets can miss the signs. Infected sheep often display mild symptoms, if any, and suffer from other conditions that may be confused with FMD.[49]

---

[47] For a diary account of an away posting during lambing, see Buckle (2001).

[48] For discussion of the complex character of partial connections, see Strathern (1991).

[49] Foot and Mouth Disease 2001: Lessons to be Learned Inquiry (2002, 49).

The sheep is enacted in a specific way here (i.e., as a potential host for the virus) but this does not mean it is passive. Instead, in these sentences the action shifts. First, FMD (foot and mouth disease) is called upon as an actor: it *is* difficult to diagnose. Then farmers and vets do something: they *miss* the signs. And then, finally, the infected sheep themselves appear to be stubborn and specific as well. They *display* symptoms in a mild way only and *suffer* from other conditions that may be confused with FMD. So the fact that it is hard to establish whether or not a specific sheep hosts the virus is the result of joint action. Disease, farmers, vets and sheep, all make a difference to the end result. To attribute all the agency to just one of these actors would be to miss the point.

It is not even easy to separate out the contribution made by each individual actor to what is jointly being done. Sheep, we just saw, may either display symptoms or hide the virus without showing it. But when farmers and vets want to act in their turn, they need the sheep to collaborate. If farmers and vets want to try to *not* miss the signs of the disease the sheep needs to open its mouth. So when they act themselves, again they do not act alone. As a farmer writes:

> He [the vet] had a good look round and all cattle, pigs, and sheep were given a clean bill of health even though one of last year's pet lambs trotted over with froth all round it's mouth! Goodness knows what it had been up to but it must have been a bit surprised at the reaction it got: grabbed, turned on its back and three people trying to look into its mouth. It may decide to skip the pleasantries next time I go into the shed.[50]

If a sheep is grabbed and turned on its back then it is being done to, but even a sheep brutalised in this way is not passive. To begin with, it is *surprised*. And, let us face it, it also *allows* the human beings around it to act in this brutal way. It does not, for instance, respond by biting them, as it would if it were a zebra.[51] What is more: this sheep is active in remembering its man-handling. It may have been a pet lamb last year, a lamb fed with bottles that learned to trust human beings, but next time the farmer enters the shed, it may well *decide* to skip the pleasantries. It is not stupid!

Sheep Un/counted

Sheep do not catch foot and mouth as easily as some other species, and neither do they transmit it so fast. This means that they made difficulties for the epidemiological models and for the modellers. For we may say that in epidemiology sheep are enacted 'as inputs and outputs of calculations', but if they are to be enacted in this way, the calculations need to take sheep-data into account. Extrapolation from other statistics is not enough. This was a problem in 2001 because earlier foot and mouth outbreaks had not been driven by a reservoir of infection in the sheep population.

---

[50] Stockdale (2001, 223).

[51] This is why zebra never got domesticated. For discussion see Diamond (1997).

> Dr Donaldson had criticisms of the scientific basis of the mathematical models and the influence they had had on disease control policy. He strongly disagreed with the conclusions of the modellers about their forecasts for the development of the epidemic. The predicted epidemic curves had been based on parameters from the 1967/8 epidemic which had been cattle-driven;[52]

Because cattle and sheep respond to the foot and mouth virus in different ways, epidemiological models based on data about cattle were of questionable utility in 2001. Models based on pigs would have not been good enough either. Not only are their bodies different, but also their habits. A pig's body emits more virus than that of a sheep and picks it up more easily. And while sheep in Cumbria live spread out across a hillside for many months of the year and are not in especially close contact with one another, pigs are often crammed together in sheds.[53] It is not difficult to understand why the virus might spread faster among the latter than the former.

But if sheep act in their own specific way in an epidemiological model, this does not mean that they are able to force modellers to take them into account. If some models just skip them and their specificities, what can they do? And this is what happened, in March 2001. The Imperial model did not attend to what sheep do differently.[54] No amount of bleating would have made any difference to this. Sheep can act directly in various ways but not in Imperial College and its computers. They need someone to transport their actions and someone to open Imperial's doors for them. As it is elsewhere, so too it is for epidemiological models: sheep do not act alone. They only act if others collaborate with them.[55]

In March 2001 the data fed into the models, probably adding new infections in cattle and pigs to old infections in sheep, arrived at an overinflated and excessively alarmist conclusion about rates of infection. And policy was based on the latter. On March 27th, less than two weeks after the 3 km cull of sheep in Cumbria was put in place, a general and much more severe policy of slaughtering all animals on 'contiguous premises' was announced. If the 3 km cull had met with resistance, this new policy was even more controversial. Many farmers thought it was quite unnecessary. In an effort to avoid compliance, they began to hide their sheep and to avoid all contact with officialdom.

> We heard of vets being paid to ring round their clients and 'spy' on who might have animals left and might be trying to hide them. I got to a stage where I would not answer the phone unless I recognised the number on the caller display. Through all of this, our sheep, 'the girls', lived locked in the barn. The gate to our property was permanently locked.[56]

---

52 Donaldson (2002).

53 Dring (2001).

54 Scudamore (2002, 13).

55 For a fascinating account of the need for collaboration in action, see Callon and Rabeharisoa (2004).

56 Taylor (2004).

One might say that the Cumbrian sheep along with their farmers took revenge for being ignored by the most influential epidemiological modellers. The farming specificities on the ground did not match the policies based on the models. Outrage led to clashes and scarcely disguised disobedience. Indeed, in practice the cull was often voluntary and the sheep handed over for slaughter were from farmers who also had cattle which they wished to protect or from those who most wanted or needed compensation.[57] So things did not turn out in the way the government had hoped or expected. Having been ignored in the dominant epidemiological model, the sheep found other ways of acting. And thus, in their own fluid way, they became politically dangerous.[58]

The Price of Sheep

Enacted economically, sheep yield a price. However, the particular price paid for a particular sheep does not depend on that sheep alone. As is obvious, it reflects whatever comes together at the point of buying and selling in, or indeed outside, the market.[59] For Cumbrian sheep in March 2001, it was partly a matter of taste (why do consumers in Spain, Italy and the South of France like the light lamb of the northern fells, while British customers do not?[60]). It was partly a matter of international regulation (since no exports were possible in 2001, the price of lamb fell for farmers in Britain, while, as a result of the consequent shortages, they rose in continental Europe.[61]) It was partly a matter of European policy. (The CAP headage payments, calculated on the basis of market prices in continental Europe, were unusually low precisely because those market prices were high.[62]) It was partly a matter of the exchange rate between the euro and the pound (the CAP headage payment is calculated in euros, but the pound was high in 2001, so the payment was low). It was partly a matter of domestic regulation. (On 2nd March, after a complete standstill, it was announced that animals could be taken under licence directly to the abattoir but since the number of hours permitted for drivers was limited, in practice the abattoir needed to be within 160 miles of the farm.[63])

So, it is not 'the individual sheep' in and of itself that yields a price. Instead, it is an embedded sheep. Earlier we said that a sheep does not act alone but that

[57] Heaton (2002).

[58] For agency understood as fluidity, see de Laet and Mol (2000).

[59] There is fine work on the specificities of markets, and the enactment of commodities in Callon (1998b). See, in particular, his (1998a).

[60] House of Commons (2001, 90 WH).

[61] House of Commons (2001, 94 WH).

[62] House of Commons (2001, 94 WH).

[63] See Peter Atkinson MP at House of Commons (2001, 91 WH) and National Audit Office (2002, 102).

action moves around in a fluid way. Then we added that if a sheep *may* act (say, in a model), this does not mean that it will actually do so. It may well require the collaboration of other actors. Now, we add that the other elements of the web in which it is embedded, not only allows a sheep to act (or not), but also influences *what* exactly it may do. They afford it in specific ways. In March 2001, consumer taste, international regulations, European policy, exchange rates and abattoir distance, to name just a few of the relevant (f)actors jointly afforded a Cumbrian sheep with a price.

This is *not* to say that these (f)actors *determined* its price. There were so many of them that their activities interfered with one another in unpredictable ways. And, in any case, the sheep itself also made a difference. It acted too. For instance, if it died before it came to market (of disease or at the hands of slaughter men as part of the cull), other factors came into play. A carcass might yield nothing at all or it might attract a (better or worse) compensation payment.

> The average compensation values for sheep rose from an initial £100 to £300 in July and then also declined. A standard rate card was introduced in March 2001. To encourage its use the standard rates were based on the upper quartile of market prices before the February outbreak. The rates ranged from ... £32 to £150 for sheep.[64]

Enacted as an economic entity, a sheep acts in an economic way. It does something that fits into economic calculations: it yields a price. But since it does not do so alone, it cannot begin to control the level of the price. Action is very different from control.

A Flock on the Hill

Sheep have cohabited with humans for millennia[65] and they have done so in many different modes. Shepherds no longer live in the Lakeland hills with their summer sheep, but even so farming practice in Cumbria still enacts sheep as animals in need of care. Yet even as they are being cared for, sheep also take care of themselves. Of course they actively graze and even if some lambs are bottle-fed at the farm, most are nursed by the ewes. However, the most striking self-care activity of Cumbrian sheep is that even though they wander across the hills without a shepherd, they know where to go.[66]

> The heafed or hefted flocks of hill sheep, to the best of my knowledge, are unique to Great Britain and Ireland. From my own experience, having established a new flock on a hill, it takes three to five years for the flock to learn their boundaries. During this time sheep are lost through straying, from drowning in streams and bogs because they don't

[64] Foot and Mouth Disease 2001: Lessons to be Learned Inquiry (2002, 132).

[65] At least nine to ten thousand years. See Clutton-Brock (1999, 74).

[66] The former care of shepherds may well lie at the origins of hefting that can thus be understood as a form of self-care. See Smart (2001).

know the safe passages and buried in snow because they haven't learned the good shelter areas.[67]

This particular Cumbrian farmer adds that a flock needs at least five years to become knowledgeable enough to be safe. So knowing how not to wander does not come easy to Cumbrian sheep. But the fact that they do had far reaching consequences in March 2001. The extraordinary ability of flocks of hefted sheep to learn about the landscape, fed into the next change in policy:

Hefted sheep result from years of careful breeding during which knowledge of the parcel of land, the 'heft', is passed on from ewe to ewe lambs. As part of the policy adjustments announced on 26 April, rare breeds and hefted sheep were exempted from the contiguous cull provided that strict biosecurity was maintained.[68]

This tells us that how hefted sheep act in relation to the fells was reflected back in what was being done to them. Not directly, for others had to make the arguments, but even so the actions of the sheep made a difference to government policy.

Policy was decided in London, so this was a case of acting at a distance.[69] The point is not that the sheep *sought* to affect others far away or that they tried to protect themselves but that they did so. Hefted sheep are influential far beyond their own immediate reach in other ways as well. This is because they do not only *know* the landscape they live in but also actively *interfere* with it:

Dr. Bardgett of Lancaster University has carried out research into what happens when the grazing animals are removed from a hill. Within ten years it is fast reverting to scrub. First the dominant grasses will swamp the sward. Bracken will encroach even faster. Gorse, and in some cases brambles, will seed and spread unchecked. Seeds of Silver Birch, Ash, Sycamore, Willow and Alder carry on the wind and will germinate and grow unmolested. Rare plants, that thrive because of and not despite the grazing pressure, will disappear.[70]

So, hefted sheep helped to make and maintain the Lake District landscape. And this is the original landscape of the English romantic imagination. Largely unwooded and largely unfenced, with its sudden open vistas of distant valleys and lakes, it is kept in this state in part by the actions of the hefted sheep as they graze evenly across its slopes. But this implied that the romantic imagination was under threat in 2001 and doubly so. In the short run, the fells were out of bounds for walkers. The fear was that the latter might spread the virus. In the long run, there was the danger that denuded of its hefted flocks, the landscape would start to change. That scrub would start to grow and trees and that if sheep returned at all, fences would be needed:[71]

---

[67] Mawdsley (2001).

[68] Foot and Mouth Disease 2001: Lessons to be Learned Inquiry (2002, 128).

[69] On action at a distance, see Latour (1987).

[70] Mawdsley (2001).

[71] And indeed in certain areas of the fells, for instance around the Old Man of Coniston and Wrynose Pass, there are now fences to keep the sheep in place. The hope is that they can be removed once the sheep have learned where they may safely graze.

> For hundreds of thousands, perhaps millions, rambling is an escape from regimentation. They have lost the right to roam. They are being told that the slaughter of Herdwick sheep will transform the open fells of the Lake District into a scrubland no one who has loved Wordsworth or followed a Wainwright guide will recognise – and that the Peak District could follow.[72]

Without the action of its knowledgeable flocks of sheep, the open fells would disappear and with this the possibility of a specific kind of romantic 'escape from regimentation'. The sheep, to be sure, know little of this as they teach their lambs where and where not to go. They graze. So far as we know, they care little about nature lovers, walking boots and Wainwright guides. Yet their actions inform the romantic imagination and its activities which tells us that even in order to act at a distance, one does not need to be a strategist.

What Is Done

We set out to answer the question: 'do sheep act?' And we have suggested that they do. But along the way we have interfered with and changed the terms of the question. Both the words 'sheep' and 'act' now mean something different.

A sheep, to begin there, is not a figure given with the space-time co-ordinates that we tend to use for geographical and historical localising. We need to say that the sheep we are talking about was found in Cumbria in March 2001. But this is just a beginning, for then there are practices, a whole variety of them, to be examined. So we have tried to show that veterinary, epidemiological, economic and farming practices each enacted 'sheep' in different ways which tells us that what a 'sheep' *is* can only be known if we explore these practices – that it is not given outside them. But this is not to say that exploring a few practices in a few pages gives us an exhaustive answer to the question 'what is sheep?' Our inquiries are necessarily partial. We might say a lot more about veterinary, epidemiological, economic and farming practices and we might also go on to investigate practices to do with tourism, animal rights, slaughter and the disposal of the carcasses. The list is endless. The reality of an entity is never exhausted. Imagine it as a fractal: if you magnify a fragment you discover an image that is as complex as the first one. And it is the same if you shift your attention to another fragment.[73]

So, a sheep-enacted does not *exist* all by itself and neither does it *act* alone. Indeed, an actor-enacted acts in collaboration with others to such an extent that it is not always clear who is doing what. Action moves around. It is like a viscous fluid. What each actor does also depends on its co-actors, on whether they allow it to act and on what they allow it to do, on rules and regulations. But this is not to say that an actor-enacted is determined by its surroundings. It has

---

[72] Cohen (2001).

[73] This argument is explored in Strathern (1991).

its own stubbornness and specificities: it is full of surprises. So the difference an actor makes is not predictable. Indeed, on the contrary: what actors-enacted do is essentially indeterminate. So much comes together in the collaborative webs of complex practice. How might one begin to know beforehand how it adds up, how the various tugs and pulls intersect and interfere with one another?

So, the actor-enacted is complex. And that complexity only increases if we also attend explicitly to the *normativity* of all the activities in which it is involved. We have not said a lot about this but it is obvious that stories about foot and mouth disease are far from neutral. The disease brings dis-ease. It also affects what is appreciated. Take the mundane activities of farming. In normal times these are hard work. Periods of grind. Lambing in the middle of the night. Hand feeding a number of lambs. But suddenly, when those activities are no longer needed, they look different:

> Since the start of lambing in early January I had been waiting for the day when the kitchen windowsill would be free from bottles, teats, jugs and a box of milk powder to feed orphan lambs. When that day came, how I wished them back again.[74]

And foot and mouth was terrible for many:

> What do you say [says an MP] to men, big men with hands like shovels and hearts like an ox who are broken down on the phone? They say, 'Eeh! I canna talk to you any more lad. Can you speak to the wife?'[75]

But this does not mean that nothing good happened in Cumbria in March 2001. Again we discover complexities. In every specific context and at every moment, there are ways of handling things that are better and ways that are worse.. Goods and bads are being done at every level of scale. They also get mixed up and what is more, there are ambivalences.[76] So that, for instance, even at the worst moments for farmers, when their animals were slaughtered, there were nuances. Farmers sent bunches of flowers to the vets,[77] and good slaughter men could still be recognised as good slaughter men:

> It sounds like it is very tense work, arriving on farms and not knowing what the handling facilities are going to be like, or the people – vets and farmers. But George and his team obviously feel a sense of responsibility to both farmer and livestock and have been managing to keep the stress on the animals to a minimum.[78]

Here, the slaughter men are appreciated for their 'sense of responsibility'. But often the normativities of action have little to do with intentions. They are unintended because they were not predicted or because where the action was being taken no one had thought the consequences through:[79]

---

[74] Mercer (2002, 77).

[75] Maclean (2001, 55).

[76] On ambivalence, see Singleton (1996).

[77] Frost-Pennington (2001, 7).

[78] Stockdale (2001, 114).

[79] The classic essay on unintended consequences comes from Merton (1957, 60–69).

> structural changes in the sheep industry which over a period of years have resulted in an increase in the size of the national flock, a reduction in the farm labour force resulting in greater reliance on shared or contracted labour and the fact that >50% of livestock holdings have sheep on them at some time of the year.[80]

An increase in the size of the national flock is good for farmers if it increases their incomes. But epidemiologically it is bad: that increase in size, together with the trading that goes with it, increases the likelihood of an epizootic. Indeed it can be argued that one of the contributory factors of the size of the 2001 outbreak was the sheer scale of the (partially CAP related) trading in sheep in late February.[81] An increase in the number of sheep is also bad for the landscape:

> Environmentalists argue that the highlands bear too much grazing, in part because of the CAP headage payments for ewes.[82]

The landscape depends on the sheep – but if there are too many of them, it gets destroyed. Activities have complex and often unpredictable effects. And the interfering webs that make up assemblages lead to surprises, too. All of which means that what emerges is hard to predict. For assemblages, like actors, are *creative*. They have novel effects and they make new things. However, to say that they make new things tells us nothing about the desirability of those new things. Indeterminacy and novelty are indifferent to the human condition – or, for that matter, to that of the sheep or the hills. But this creativity leaves us with a possible definition of an actor. We might say that an actor is a *moment of indeterminacy* that generates events and situations. It does this together with other actors that enact it and that it, in its turn, enacts. And it does so for better, or for worse, or both. Or, then again, we might say that an actor is the site where we situate what surprised us *post hoc* when we tell stories about events and situations. It is the creative limit where our stories stop.

But if we define the actor in this way, then this suggests that determining whether this or that is or is not an actor is of secondary importance. More interesting than the fact that things may act is what they do. Anything is, or might be, or might be said to be, an actor. So the point is not *who* has done it.[83] Instead, what become more urgent are questions about *what* is happening. What do actors *do*? How are they creative? How do their underdetermined activities help to create or to destroy? What are the possibilities that they condition? Or, to speak as a walker in the Lake District hills: where does this path come from and where might it lead?

---

80 Department for Environment Food and Rural Affairs (2002, 2).

81 Foot and Mouth Disease 2001: Lessons to be Learned Inquiry (2002, 30).

82 Policy Commission on the Future of Farming and Food (2002, 73).

83 No wonder that even in the most beautiful social science 'whodunit', the author, Latour cannot find the answer to the question: 'who killed Aramis' (see Latour (1996)).

References

Akrich, M., 1992, The De-Scription of Technical Objects. In *Shaping Technology, Building Society: Studies in Sociotechnical Change*, edited by E. Bijker and J. Law, pp. 205–224. MIT Press, Cambridge, MA.

Ashworth, S., Palmer, M., and Northen, H., 2000, *An Economic Evaluation of the Sheepmeat Regime as Applied in the United Kingdom*, Auchincruive, Ayr: SAC, also available at http://statistics.defra.gov.uk/esg/evaluation/shpmtreg/.

Bingham, N., 2006, Bees, Butterflies and Bacteria: Biotechnology and the Politics of Nonhuman Friendship. *Environment and Planning A*, 38(3): 483–498.

Buckle, M., 2001, No title. In *Foot and Mouth: Heart and Soul. A Collection of Personal Accounts of the Foot and Mouth Outbreak in Cumbria 2001*, edited by C. Graham, pp. 34–47. BBC Radio Cumbria, Carlisle.

Callon, M., 1986, Some Elements of a Sociology of Translation: Domestication of the Scallops and the Fishermen of Saint Brieuc Bay. In *Power, Action and Belief: a new Sociology of Knowledge?*, edited by J. Law, pp. 196–233. Sociological Review Monograph, 32, Routledge and Kegan Paul, London.

Callon, M., 1998a, Introduction: the Embeddedness of Economic Markets in Economics. In *The Laws of the Markets*, edited by M. Callon, pp. 1–57. Blackwell and the Sociological Review, Oxford and Keele.

Callon, M. (ed.), 1998b, *The Laws of the Markets*, Blackwell and the Sociological Review, Oxford and Keele.

Callon, M., and Rabeharisoa, V., 2004, Gino's Lesson on Humanity, *Economy and Society* 33: 1–27.

Clutton-Brock, J., 1999, *A Natural History of Domesticated Animals*. Cambridge University Press, Cambridge.

Cohen, N., 2001, Well, Goodbye Dolly. *The Observer*, 8th April, 2001, also available at http://politics.guardian.co.uk/footandmouth/comment/0,9236,470361,00.html.

Cumbria Foot and Mouth Disease Inquiry, 2002, Report. Cumbria Foot and Mouth Disease Task Force, Carlisle. http://web.archive.org/web/20021016132125/http://www.cumbria.gov.uk/news/footandmouth/inquiry/f&m_inquiry_report.pdf, updated 16th October, 2002, (accessed 29th August, 2003).

de Laet, M., and Mol, A., 2000, The Zimbabwe Bush Pump: Mechanics of a Fluid Technology, *Social Studies of Science* 30(2): 225–263.

Department for Environment Food and Rural Affairs, 2002, Origin of the UK Foot and Mouth Disease Epidemic in 2001. London, http://www.defra.gov.uk/corporate/inquiries/lessons/fmdorigins.pdf, updated June 2002, (accessed 29th August, 2003).

Diamond, J. M., 1997, *Guns, Germs and Steel : a Short History of Everybody for the Last 13,000 Years*. Jonathan Cape, London.

Donaldson, A., 2002, Note of Meeting 8 April 2002. Lessons to be Learned Inquiry, London. http://archive.cabinetoffice.gov.uk/fmd/fmd_report/documents/B-MeetingNotes/Institute%20of%20Animal%20Health%20-%20Dr%20Alex%20Donaldson.pdf, (accessed 28 February 2006).

Dring, J., 2001, My Involvement with the Waughs. DEFRA, London, http://www.defra.gov.uk/footandmouth/pdf/dringstatement.pdf, updated 16th March 2004, (accessed 9th August 2004).

Foot and Mouth Disease 2001, Lessons to be Learned Inquiry, 2002, Report, London: London, The Stationery Office, http://213.121.214.218/fmd/report/index.htm, (accessed 28th August, 2003).

Frost-Pennington, P., 2001, Into the Valleys of Death. In *Foot and Mouth: Heart and Soul. A Collection of Personal Accounts of the Foot and Mouth Outbreak in Cumbria 2001*, edited by C. Graham, pp. 7–8. BBC Radio Cumbria, Carlisle.

Haraway, D. J., 1991, Situated Knowledges: The Science Question in Feminism and the Privilege of Partial Perspective. In *Simians, Cyborgs and Women: the Reinvention of Nature,* edited by D. Haraway, pp. 183–201. Free Association Books, London, also available at http://www.hsph.harvard.edu/rt21/concepts/HARAWAY.html.

Haraway, D. J., 2003, *The Companion Species Manifesto: Dogs, People, and Significant Otherness.* Prickly Paradigm Press, Chicago.

Heaton, S., 2002, Cumbria Public Meeting. Lessons to Be Learned Inquiry, London, http://archive.cabinetoffice.gov.uk/fmd/fmd_report/documents/A-Submissions/Ref%20414.pdf, (accessed 21 February 2006).

Hinchliffe, S., 2001, Indeterminacy In-Decisions: Science, Policy and Politics in the BSE Crisis, *Transactions of the Institute of British Geographers* 26: 182–204.

House of Commons, 2001, Upland Farming, *Hansard* 372: (23), Cols 96 WH, (18 July 2001), also available at http://www.publications.parliament.uk/pa/cm200102/cmhansrd/vo010718/halltext/10718 h02.htm#10718 h02_spnew4.

House of Commons Select Committee on Agriculture, 2001, *Minutes of Evidence: Examination of Witnesses (Questions* 480–499*), Wednesday 25 April, 2001,* The Stationery Office, London, also available at http://www.publications.parliament.uk/pa/cm200001/cmselect/cmagric/363/1042508.htm.

Jackson, H., 2001, Foot and Mouth. In *Foot and Mouth: Heart and Soul. A Collection of Personal Accounts of the Foot and Mouth Outbreak in Cumbria 2001,* edited by C. Graham, pp. 108–109. BBC Radio Cumbria, Carlisle.

Kao, R. R., 2002, The Role of Mathematical Modelling in the Control of the 2001 FMD Epidemic in the UK, *Trends in Microbiology* 10(6): 279–286.

Latour, B., 1987, *Science in Action: How to Follow Scientists and Engineers Through Society.* Open University Press, Milton Keynes.

Latour, B., 1988, *Irréductions, published with The Pasteurisation of France.* Harvard, Cambridge, Mass.

Latour, B., 1996, *Aramis, or the Love of Technology,* MIT Press, Cambridge, Mass.

Law, J., 2002, *Aircraft Stories: Decentering the Object in Technoscience*, Duke University Press, Durham, N.Ca.

Law, J., 2003, *Traduction/Trahison: Notes on ANT*, University of Oslo, Oslo, TMV Working Paper, 106, also available at www.lancs.ac.uk/fass/sociology/papers/law-traduction-trahison.pdf; 1997.

Maclean, D., 2001, Foot and Mouth. In *Foot and Mouth: Heart and Soul. A Collection of Personal Accounts of the Foot and Mouth Outbreak in Cumbria 2001,* edited by C. Graham, pp. 53–57. BBC Radio Cumbria, Carlisle.

Mawdsley, R., 2001, Diary. *Heart of Cumbria,* http://web.archive.org/web/20020123225817/www.heartofcumbria.com/diaries/mawdsley1.htm, (accessed 26th July 2004).

Mercer, I., 2002, Crisis and Opportunity: Devon Foot and Mouth Inquiry 2001. Tiverton, Devon Books, http://www.devon.gov.uk/fminquiry/finalreport/, (accessed 28th August, 2003).

Merton, R. K., 1957, *Social Theory and Social Structure.* Free Press, New York.

Mol, A., 2002, *The Body Multiple: Ontology in Medical Practice*, Duke University Press, Durham, N. Ca., and London:

Mol, A., 2006, *De Logica van het Zorgen: Actieve Patiënten en de Grenzen van her Kiezen*, Amsterdam: Van Gennep; (forthcoming in English as *The Logic of Care: Active Patients and the Limits of Choice*).

National Audit Office, 2002, The 2001 Outbreak of Foot and Mouth Disease. National Audit Office, London, HC 939, http://www.nao.gov.uk/publications/nao_reports/01-02/0102939.pdf, updated 21st June 2002, (accessed 15th September 2003).

Policy Commission on the Future of Farming and Food, 2002, *Farming and Food: A Sustainable Future.* Cabinet Office, London, also available at http://www.cabinet-office.gov.uk/farming/pdf/PC%20Report2.pdf.

Scudamore, J. M., 2002, Note of Meeting 10 April 2002. Lessons to Be Learned Inquiry, London. http://archive.cabinetoffice.gov.uk/fmd/fmd_report/documents/B-MeetingNotes/Jim%20Scudamore%20First%20 Meeting.pdf, (accessed 26 February 2006).

Shannon, D., 2002, Note of a Meeting on 8 April 2002 with Dr David Shannon, retired Chief Scientist, DEFRA. Lessons to Be Learned Inquiry, London. http://archive.cabinetoffice.gov.uk/fmd/fmd_report/documents/B-MeetingNotes/Dr%20 David%20Shannon.pdf, (accessed 24 February 2006).

Singleton, V., 1996, Feminism, Sociology of Scientific Knowledge and Postmodernism: Politics, Theory and Me, *Social Studies of Science* 26: 445–468.

Smart, K., 2001, Herdwick Sheep – the Irreplaceable 'Wild' Sheep who are the Lake District's Gardeners. News and Star, 24 March 2001, also available at http://www.visitcumbria.com/herdwick.htm.

Stockdale, M., 2001, Foot and Mouth. In *Foot and Mouth: Heart and Soul. A Collection of Personal Accounts of the Foot and Mouth Outbreak in Cumbria 2001*, edited by C. Graham, pp. 112–117. BBC Radio Cumbria, Carlisle.

Strathern, M., 1991, *Partial Connections*. Rowman and Littlefield, Savage Maryland.

Taylor, C., 2004, Coleen's Story. *Farmtalking*, http://web.archive.org/web/20020411165451/http://www.farmtalking.com/coleenstory.html, updated 19th July, 2004, (accessed 5th January 2007).

Taylor, N., 2003, *Review of the use of Models in Informing Disease Control Policy. Development and Adjustment*, Reading: Veterinary Epidemiology and Economics Research Unit (VEERU), School of Agriculture, Policy and Development, The University of Reading, also available at http://www.defra.gov.uk/science/documents/publications/2003/UseofModelsinDiseaseControlPolicy.pdf.

The Royal Society, 2002, Infectious Diseases in Livestock. Scientific Questions Relating to the Transmission, Prevention and Control of Epidemic Outbreaks of Infectious Disease in Livestock in Great Britain. Royal Society of London, London. http://www.royalsoc.ac.uk/inquiry/, (accessed 29th August 2003).

Thévenot, L., 2001, Which Road to Follow? The Moral Complexity of an 'Equipped' Humanity. In *Complexities: Social Studies of Knowledge Practices*, edited by J. Law and A. Mol. Duke University Press, Durham, N.Ca

Chapter 5
Non-Human Agencies: Trees in Place and Time

Owain Jones and Paul Cloke

'Living things are restless' (Tudge: 2005: 16)

Introduction

Moby sings "we are all made of stars". He does this on the basis that all the complex atomic elements which go to make up the earth, and life on the earth, were produced by the nuclear fusion of distant, ancient stars, that exploded, scattering the elements into vast stellar dust/gas clouds which, eons later, got compressed and formed planets, then the life on them. Relational materiality can be problematically all-embracing and far-reaching, but it is quite clear that the social aspect is thoroughly dependent on the life – making capacities of a whole range of natural processes which are articulated through various forms, flows and exchanges of energy and matter/materiality. We depend utterly for life on our own sun, on the biosphere systems of atmospheric regulation, on gravity, on the magnetic field of the earth (maintained by the fluid molten iron core of the earth), on the billions upon billions of microbes both in our bodies and in the soil, on insects, on plants, on trees – this could be a very long and detailed list.

Social life is bound into all these almost untraceably complex, intersecting, far-reaching space-time material patterns, but this is not a fixed binding. Social life can, in turn alter the processes into which it is woven, at both the local and global scales. Indeed, the capacity of humans to act creatively – a basic definition of agency – often leads to the view that we are the only force in the world equipped with agency. We argue, along with others, that this denial of non-human materiality is both deluded and potentially dangerous. Non-human agencies not only co-constitute the contexts of life, but they also frequently reconstitute the fabrics of day-to-day life and the places and spaces in which it is lived. Bodies, houses, cities, offices, countryside and so on should all be viewed as contributing to human relations in

O. Jones
Countryside and Community Research Institute, Dunholme Villa, The Park, Cheltenham, UK
e-mail: ojones@glos.ac.uk

C. Knappett, L. Malafouris (eds.), *Material Agency*,
DOI: 10.1007/978-0-387-74711-8_5,

myriad ways. On the basis of this realisation, a range of approaches are now reopening the question of non-human agency, relational agency, and, not least, the agency of materiality. In this chapter, we review some of these approaches, focussing on new conceptualisations of place and time in human geography that seek to re-embrace the agency of non-humans and the politics and ethics which are affected by such agency. Our empirical context for this exploration is a research project in which we examined the agency of trees in different case study places.

Taking Nature and Materiality Seriously

In geography and beyond, 'nature has remained a largely undifferentiated concept, its constituent parts rarely theorized separately' (Wolch and Emel, 1998). As David Harvey (1996: 183) puts it, there has been a tendency to 'homogenize the category "nature" (and [to] discuss its social meaning and constitution as a unitary category) when it should be regarded as intensely internally variegated – an unparalleled field of difference'. In this section, we review recent discussions of the interconnectivity of human and natural worlds, noting how the notion of agency has been translated across a previously disabling binary distinction between human and non-human. It is clear from any such review that the understanding of agency depends significantly on the precise theoretical questions being posed. For example, the development of Actor Network Theory (ANT), (Callon, 1986, 1991; Callon and Law, 1995; Latour, 1993; Law, 1994 – see also Murdoch, 1995, 1997; Thrift, 1996; Whatmore, 1999, 2002), has explicitly valorised a perspective on agency which accentuates the *relational*, subjugating the specific importance of individual actors within networks, in order to focus on the multiplicity of mutually constitutive and positioning 'actants' which together serve to hybridise agency. ANT, therefore, has both emphasised the 'non-humanness' of agency and yet declined to recognise it per se because of ANT's overall theoretical project of deconstructing the antimony of nature and society.

Our concern with the *agency* of non-human actants such as trees leads us to ask some different but related questions. In one sense, we are seeking to give others 'their due' (Thrift, 1996: 26). While relational agency is crucial to the project of hybridity, there remain very significant questions about the potential contribution brought by non-human agents such as trees to hybrid relations. Our discussion of varying types of agency (Jones and Cloke, 2002) confirms that trees act upon as well as being acted upon. As part of our research into the interconnections between trees and places, we proposed four ways in which trees – and other non-humans – might be regarded as having "agency":

1. *Agency as routine action*: trees are associated with a series of ongoing processes of existence which enable them to grow, reproduce, bear fruit, spread, colonise and so on. Although such processes may be associated with human intervention (e.g., planting, pruning, cutting down), the tree clearly transcends the passive role often allocated to nature's subjects.

2. *Agency as transformative action*: trees can be seen to make new directions and formations. They are active in the creation and folding fields of relations, which in turn is often bound up with the transformation of places. Trees can act autonomously in seeding themselves and growing in unexpected places and in unexpected forms and when remixed with the social aspect, these actions can have creative transformative effects.
3. *Agency as purposive action*: intentionality is a key threshold by which agency is often limited to the social realm. Indeed, ascribing intentionality to non-human agents can lead to dangerous forms of reductionist essentialism. However, non-humans do exercise a kind of purposive agency, for example, in the way that trees are able to influence future courses of action; their DNA clearly entertains a plan which purposes particular forms of being and becoming – an implicit blueprint with instructions for its construction and physiological functioning (Gordon, 1997). The skill of trees is, then, to have a means of executing embedded purposeful agency, which is capable of exploiting myriad circumstances and thereby influencing place production.
4. *Agency as non-reflexive action*: the socio-ecological world exhibits significant creativity and creative potentials and non-agents such as trees participate fully in creative being and becoming. In particular, trees have a capacity to engender affective and emotional responses from the humans who dwell amongst them – to contribute to the haunting of place via exchanges between the visible present and the starkly absent in the multiple and incomplete becoming of agency.

In identifying these four strands, our argument is not that trees possess the particular and extraordinary capabilities of humans in these respects but that they do possess very significant forms of active agency, which have usually been assumed to exist only in the human realm.

This view of non-human others as active agents is important for a wide range of ontological, epistemological, political and ethical reasons. However, it also begs further questions about a purely relational view of nature-society in which processes, flows, fluxes – and the networks which are their fabric – render notions of autonomous separated individuals as untenable. To start with, we need to question the extent to which what we might commonly consider to be individual entities (such as trees) are destabilised and re-assembled in hybrid bodies or networks. More generally, much of the illustration and application of this hybridity seems to have been biased towards technological rather than organic non-human entities – a manoeuvre which somehow makes it easier to deny the specific non-human contribution to hybrid agency (and perhaps continues, implicitly, to value human authority in the shaping of hybridity and relational networks). If, as Whatmore (1999: 26) suggests, agency is 'a relational achievement, involving the creative presence of organic beings, technological devices and discursive codes', then what is the nature of that 'creativity' that these 'beings' bring to the relational process?

Both in our own work (Jones and Cloke, 2002) and in that of far more eminent theoreticians (e.g., Latour, 2004), it is pointed out that the term 'agency' needs to be disaggregated in order to account for the differing ways in which the global population of things can act creatively. As Latour (2005: 226) says:

> there might exist many metaphysical shades between full causality and sheer non-existence [in terms of agency]: things might authorise, allow, afford, encourage, permit, suggest, influence, block, render possible, forbid and so on. [ ] No *science of the social can even begin if the question of who and what participates in action is not first opened up, even though it might mean letting elements enter, that, for lack of a better term, we call nonhumans.* (emphasis added)

In relation to trees, (and other non-human 'beings') the failure to articulate non-human agency within *its own ecological time-scales* as well as in its own places has made it difficult to grasp the notion of non-human agency within extant and more anthropocentric views of agency.

Reviewing the World: Replacing the Human

> Agency [ ] is the other side to structure, the materiality of the social world which is to be changed by the human subject. In this approach, agency is a characteristic of a consciousness adequate to the social world, and which thus puts change into motion (Game, 1991: 12).

> It has been commonplace in social theory to ignore the specific 'agency' and materiality of nature, or, where that agency has been admitted, to conceive of it within the disabling binary logic that has for so long organised modern thought (Fitzsimmons and Goodman, 1998: 194).

Over recent years there has been a significant re-viewing of the world by social theorists who argue that previous ideas of 'agency' have been far too narrow. By defining agency in terms of the 'human subject', previous social theory had tended to ignore the agency and materiality of nature, leading to the social and scientific construction of nature in human terms. In Latour's (1993: 138) words, 'modern humanists are reductionist because they seek to attribute action to a small number of powers, leaving the rest of the world with nothing but simple mute forces'. The major stumbling block to acknowledging agency beyond the human sphere has been the attribution of apparently unique characteristics of purpose and communication to human beings. Callon and Law (1995: 491) suggest that in contemporary Western cultures 'something is treated as an agent, or at any rate, as a candidate for agency, if it performs, or might perform, two great classes of condition: intentionality and language use'. These agency 'thresholds' have previously dictated that agency is regarded as a human process. Whatmore (1999: 29) highlights that it is 'a refusal to equate agency (the capacity to act or have effects) with intentionality, premised on narrow linguistic competences' which has been 'the point of greatest tension between ANT and conventional social theories'.

This restricted view of meaningful agency has become untenable in many different streams of social theory, in which it is argued that differing forms of agency need to be recognised and understood. We recognise four such streams to be important in this context. First, a significant theoretical prompt for non-human agency has come from eco-feminism. Warren (1994: 4), for example, suggests that 'acknowledgement of the world's active agency seems to be necessary to the deconstructive process of dismantling totalising and essentialising discourse and it is something eco-feminist philosophy can and does do well'. In particular, it is Donna Haraway's seminal writings that have brought eco-feminist concerns to bear in the sociology of scientific knowledge. She has concluded that 'nature' is a multidimensional tangle of the political, economic, technical, cultural, mythic and organic which 'collapse into each other in a knot of extraordinary density' (Haraway, 1994: 63). Her work 'envisions "feminist theory [ ] with actors who come in many and wonderful forms"' (Cheney, 1994: 170–1; citing Haraway, 1988) and the agencies of nature have been spotlighted as a consequence.

A second stream of influential social theory has emerged from critical development of ideas about 'social nature' (Fitzsimmons, 1989a,b; Harrison and Burgess, 1994), in which radical geographies of political economy have become increasingly reconciled with the need to regard nature as an essential third theoretical arena in addition to society and space. The initial argument that nature cannot be (re)produced outside of social relations was quickly tempered by the equally significant argument that nature is not reducible to such social relations:

> Rather, the biological and physical dynamics of life forms and processes need to be recognised on their own terms, conceptually independent of human social agency, such that social nature represents "the outcome of a specific structure of natural/social articulation" (Benton, 1989). Following from this, such structures of natural/social articulation are spatially as well as historically specific.[ ] As a living complex of life forms and processes, natural relations are always embedded and thereby interact with, and condition, human social relations to varying extents and in different ways in specific times and spaces (Whatmore and Boucher, 1992: 168).

These theoretical trajectories are taken up and pressed by David Harvey (1996) who has lent his considerable personal standing to the task of understanding how fundamental [biophysical] conditions and processes work together with social, economic and cultural projects to create tangible historical geographies and to do so in ways which do not position the physical and the biological as banal and passive contexts within which dominant human processes occur. He too, then, argues that non-human organisms should be seen as 'active subjects' capable of 'transforming nature' and 'adapting to the ecosystem they themselves construct' (p186). And in charting this theoretical territory, he further argues that the 'artificial break between "society" and "nature" must be eroded, rendered porous, and eventually dissolved' (p192).

A third and more recent theoretical development in the re-placing of the human has come from socio-anthropological writings. For example, Macnaghten and Urry (1998) have recognised the significance of non-human agency through their development of Ingold's (1993, 2000) notions of 'dwelling' and 'taskscape'. They argue that 'landscape' and 'nature' are complex, spatial and temporal achievements and that 'relationships with what is taken to be "nature" are embodied, involving a variety of senses and that there are "physical" components of walls, textures, land, plants and so on, which partly constitute such "dwellings"' (p168). In other words, dwelling is a complex performative achievement of heterogeneous actors in relational spatial/temporal settings. Ingold (1993) illustrates these ideas in a discussion of Bruegel's painting *The Harvesters*. In this picture, an old pear tree takes centrestage as farm labourers rest and eat in its shade in between harvesting a field of grain; behind these is a view of a valley and a distant village and landscape. Ingold says of the tree

> by its presence it constitutes a particular place. The place was not there before the tree, but came into being with it. [N]o other tree has quite the same configuration of branches, diverging, bending and twisting in exactly the same way. In its present form, the tree embodies the entire history of its development from the moment it first took root. And the history consists in the unfolding relations with manifold components of its environment, including the people who have nurtured it, tilled the soil around it, pruned its branches, picked its fruit, and – as at present – use it as something to lean against. The people, in other words, are as much bound up in the life of the tree, as is the tree in the life of the people (pp. 167–8).

Our argument is that the tree is not just a passive recipient of these human interventions. Rather, it brings its own abilities and tendencies to the equation. In relational processes of fruit production, trees are not only shaped and controlled by pruning, their natures and capacities have shaped the very possibility and precise process of pruning in the first place (Cloke and Jones, 2001). More recently, Ingold (1997: 249) has suggested the need to dissolve the 'category of the social, so as to re-embed (human) relationships within the continuum of organic life'.

It is, however, the fourth theoretical strand – Actor Network Theory which has been most influential in proposing and drawing together fundamentally new ways of understanding the relations between nature and society. The development and substance of ANT have been critically reviewed elsewhere (see Braun and Castree, 1998; Murdoch, 1995, 1997; Thrift, 1996; Whatmore 1999). For our purposes here, it is sufficient to underline two aspects of ANT. First, ANT has recognised the agency of non-humans as an essential element in how the natural and the social flow into one another. For example,, Callon's (1986) classic discussion of scallop fishing treats the scallop as an active agent, rather than a passive subject of human activity. In doing so, he dismantled the existing protocols that confined agency to the social sphere and set in train a move beyond socio-biology into terrains of agency in which the human and the non-human were networked together.

The second and key contribution of ANT has been to provide what Demeritt (1994: 183) refers to as a 'new metaphor [ ] for framing nature as both a real material actor and a socially constructed object'. In fact, at least two metaphors are employed. A drawing together of Haraway's (1989, 1991) 'Cyborg's' partnerships between human and non-human actors in the mutual construction of artifactual nature – and Latour's (1993) 'hybrids' – mixtures of nature and culture – provides the metaphor for *hybrid geographies* (Whatmore, 2002). Here, agency is viewed as being spun between different actors (or 'actants') rather than manifested as solitary or unitary intent and it is decoupled from subject – object distinctions. These hybrids are then seen as mobilised and assembled into associative networks in which agency represents the collective capacity for action by humans and non-humans. The inherently collective nature of this networking – or the 'hybrid collectif' (Callon and Law, 1995) – re-theorises the notion of agency:

> The notion of the hybrid collectif implodes the inside/outside binary which discerns social action as an individual property of discrete, unitary individuals (including collective individuals). Agency is reconfigured as a relational effect generated by a network of heterogeneous, interacting components whose activity is constituted in the networks of which they form a part (Whatmore, 1999: 28).

Thus ANT has championed non-human agency whilst at the same time it has rejected the non-human/human distinction. To pose nature and society in dialectic terms, even with an inherent purpose of dissolving nature-society dualisms, is seen as an unacceptably risky strategy, which 'seems still to reproduce the dualism which we are seeking to resolve' (Fitzsimmons and Goodman, 1998: 207). By employing metaphors of hybridity and network, social theorists have sought to dismantle the binary logic which poses nature and society as opposites and which champions the social over the natural. In doing so, the further investigation of non-human agency per se becomes a necessary casualty of the manoeuvre towards networked hybridity. In effect, questions relating to the nature of non-human agency and its potential contributions to various hybrid geographies, have become regarded as theoretically dubious – a return to the bad old days of modernistic separation of nature and society.

We believe that to rule out notions of non-human agency in this way may be both to lose out on significant understandings of the ways in which non-human organisms and materials contribute to the networked agencies of hybrid collectifs and to obscure assumptions about the relative significance of different forms of hybridity. On the latter point, for example, there are strong preconceptions in social science about the hierarchy of agency potentials amongst the 'components' of hybrid collectifs:

> We tend to think that human agents are only true agents because of the essential link we feel exists between thought and action, properly so called. Thus animals are regarded as simply behaving when they 'act' because we regard them as incapable of thought and therefore somehow governed in their 'endeavours' by internal drives whose status as 'thoughts' is moot. Again, sticks and stones, considered as agents, are one step lower and 'behave' in strict accordance to the laws of nature which are regarded as determining their 'endeavours' as if themselves contributed nothing at all to what happens to them. (McPhail and Ward, 1988: 72)

Such preconceptions seem to identify animals as the 'obvious' first step in considering non-human forms of agency and indeed there has been a recent surge in 'animal studies' in recent years (Ingold, 1995, 1998; Whatmore and Thorne, 1998; Wolch and Emel, 1998) in which the agency of animals has been a focus. The level 'below' animals in McPhail and Ward's view is 'sticks and stones', where presumably, other living entities such as trees are also classified without distinction in terms of agency.

We wonder whether it is it possible that these preconceptions about differentiated 'tiers' of agency have sometimes found their way into our thinking about hybridity and networks. Although ANT theorists have been keen to destabilise anthropocentric views of agency, it does seem that actors from 'nature' – organic non-human others of one kind or another – are sometimes curiously absent from some ANT considerations. At the same time, there does seem to be an overemphasis on the relational agency of artefacts and technology. Thrift (1996: 24), for example, suggests that the 'intermediaries' which Callon and others involve in actor networks are 'usually considered to be texts, technical artifacts human beings and money': (despite Callon's early focus on scallops!). Latour's (1993) claim to have 're-established symmetry between two branches of government', is with reference to 'that of things – called science and technology – and that of human beings' (p138). Woolgar's (1991) cry of 'let's hear it for machines, for a change' is echoed in Callon and Law's (1995) concern to study machines and devices, reflecting debates about 'whether computers, robots, *or, for that matter, animals,* are "really agents" or not' (p282 our italics). Note here that while Callon and Law do gesture towards 'natural' non-humans, they venture only as 'far' as animals, which appear as something of an afterthought and the wider realm of organisms is absent.

Can it be, then, that there is a danger that the agency of organisms such as trees will be obscured or even forgotten within the ANT framework? Agency is extremely diverse in nature and locus (Harvey, 1996). We suggest that not only is there a need to move away from treating the human realm as separate, privileged and ontologically unique in terms of agency, but there is also a need to disaggregate the notion of agency itself. In the remainder of this chapter, we show how it is possible to conceptualise agency in different forms and to show how such different agencies are employed or exercised by non-humans. Trees are a fertile territory for the grounding of such conceptualisations. Collectively, they have a bewildering range of skills and/or uses and they are embedded in a plethora of relationships with humans and other non-humans. With humans they are embedded in a vast range of cultural, social, technological and economic networks as well as being highly visible in local, national and global disputes over the 'environment'. Moreover, they operate in their own ecological time, often rather different from the typical time-scales of human-centred analysis.

Places and Patterns as Entanglements of Flows of Forces and Materials

Over recent years there has been a concerted effort to rejuvenate geographical approaches to place in human geography (Thrift, 1999; Massey and Thrift, 2003; Cloke and Jones, 2001, 2004; Massey, 2005) and in particular to *apply* the above theoretical orientations in grounded studies of everyday landscapes (Jones, 2006, Massey and Thrift, 2003 Cloke and Jones, 2001; 2004). These revisionist studies are keen to jettison any notion of places as simply bounded, static, social spaces, all too easily recognised in traditional cartographies. Instead, places are recognised as temporal processes where all manner of trajectories – of people, non-humans, economies, technologies, ideas and more – come, are brought or are thrown together to assemble enduring, but also changing, formations which settle out into distinctive patterns of places, yet which are still fully networked into the wider world. As Amin and Thrift (2002: 30) put it:

> Places [ ] are best thought of not so much as enduring sites but as moments of encounter, not so much as 'presents', fixed in space and time, but as variable events; twists and fluxes of interrelation.

In this way, both the social and the natural need to be recognised as "on the move." Massey (2005) emphasises the dynamic nature of nature; it is not just a case of social flows whirling through and tangling with, more fixed grounds of nature, but nature too is on the move. If a long enough view is taken, even land itself can be seen swirling across the surface of the globe through processes of plate tectonics. Many other natural processes operate in rhythms and velocities more or less amenable to human apprehension of one kind or another, yet remain highly varied – such rhythms and velocities are recognisable, for example, in ice ages, sun spot cycles, weather patterns, planetary movements and the seasons, and corresponding life rhythms of animals and plants. To focus on our grounded example, trees can themselves be seen to be "on the move", as in waves of afforestation which followed the retreating ice sheets north across Europe at the end of the last Ice Age or in the colonisation of industrial wasteland. But they do this in their own (non-human) time and with their own skills, such as different ways of broadcasting seeds.

Thus to appreciate the agencies of nature and materiality, not only do we need to appreciate the very differing forms of beings and processes in which they are articulated, but also the very differing velocities and rhythms they might be operating in. Places are 'where spatial narratives meet up or form configurations, conjunctions of trajectories which have their own temporalities' (Massey, 2005: 139). These comings and goings in the production of place can be seen as processes of patterning (Harrison, Pyle and Thrift, 2004).

> All kinds of things can come together in the world and, in that process of encounter and settling down into at least a short-term equilibrium, they can creatively produce new kinds of organisations that are greater than the sum of their parts (p.40).

The agency of making places and patterns is relational and set in the material world as well as the social, the cultural and the economic.

Thus, a reworked, dynamic notion of place can also become a ground on which the entanglements of relationality and consequent affective politics and ethics may be approached. Massey (2005) advocates an open, non-foundational, non-purifying (or purified) notion of place that is sensitive to 'other' notions of 'other' places. These kinds of platforms are now briefly examined in three sketches of case studies from our research on tree-places.

Three Tree-Places

The histories of these three places are told in much more detail elsewhere (Cloke and Jones, 2001, 2004; Jones and Cloke, 2002). Here we offer brief sketches, providing (in a form akin to time-lapse photography) an overview of how the places have changed over time. In each case, we race through a hundred years or more, so that the actions of the trees can become clear in something like their own time.

The Square. Victoria Square in Clifton, Bristol is a small urban square, lined and punctuated by a number of mature trees, younger trees and some shrubs. Amongst these are open areas of grass, through which runs a diagonal, low-walled, paved path providing a pedestrian passage into the centre of Clifton. Outside the low stone perimeter wall, there are roads and then grand terraces of houses looking into the square.

The square was planted in the middle of the 19th century as private gardens for the terraces on two sides of the square, both built and owned by a powerful private Bristol organisation, the Merchant Venturers, who were seeking to profit from the city's expansion. The gardens were private to the houses on two sides and protected by an elaborate set of railings and lockable gates. The railings even encaged the diagonal path and the two triangular sides of the square were joined by a now filled in tunnel – such was the desire for control of access. Illustrations from the time show ornate gardens with statuary and flower beds carefully laid out before the grand architecture of the terraces.

By the turn of the 20th century, the trees that had been planted as part of the garden design had grown substantially and the gardens were proving to be highly burdensome to the Merchant Venturers in terms of maintenance costs and public controversy. The exclusivity of the gardens was being called into question by residents in other houses looking onto the square. A remarkable archive of letters shows waves of complaint and counter-complaint about the trees in the garden as they grew to full size and came to dominate the space blocking the previously open vistas of the terraces. Some objected to the extent to which the trees were shading light from houses and changing the very nature and the condition of the gardens. As a result, some trees were felled or reduced in size. This caused emotional outbursts from other residents about the mutilation and desecration of these significant natural beings. However, maintenance of the garden declined further regardless of continuing complaint. The trees kept growing.

Then, decisively, the railings were removed as part of the Second World War iron salvage programme, after which the replacement wire fences themselves became derelict, along with the garden, reflecting the decline in fortune of this once fashionable area. By the 1960 s, many of the grand houses had been turned into flats. The Merchant Venturers were by then determined to divest themselves of the ongoing burden and embarrassment of the still private, but then disreputable gardens. The development of the square for new housing was considered, but this proved impossible because of the presence of what was then a series of splendid city trees and valued (if unkempt) green space in the square. During the decades of low management, a few self-seeded or suckered trees had grown up in more abandoned corners to such a size that they joined the population of planted trees as notable individuals.

In 1967, the ownership of and responsibility for the square was transferred to the city council and the remains of the gardens were grassed over, forming a small public park of trees and open grass areas. The current form and identity of the square thus became apparently settled, conforming to, amongst other things, the popular politics of environmentalism and the need for urban green space that became prioritised in the latter decades of the 20th century. But more detailed changes continue; vast but now venerable trees have fallen or been reduced for public safety and new trees have been planted – often, but not always, with the blessing of local residents and civic groups who keep a close eye on the square and its population of trees.

Fig. 5.1 Victoria Square: then and now

The Nature and Heritage Trail: The nature trail is a small wooded hill near Bath in Somerset. It is in fact a coal tip, locally termed a 'batch', of the Camerton Colliery of the now defunct Somerset coal field. The 'batch' was formed by a century or so of mining between the 1780 s and 1898. As with many other spoil heaps in the area, the Camerton Old Pit batch was planted with a mix of conifers in the first decades of the zok century – mainly for the purposes of landscaping and slope stabilisation but also to provide a home-grown crop of pit props for the new pit in Camerton. However these were small, difficult and eventually unprofitable mines and although Camerton struggled on until the mines were nationalised in 1947, it was eventually closed in 1950, thereafter remaining in the dormant ownership of the National Coal Board. By this time, the trees planted on the batch had reached full height forming a very distinctive, closely-packed, conifer forest. A few native wild trees had self-seeded in less crowded margins and corners of the site, including a huge beech tree, which had grown up in the abandoned railway sidings alongside the batch.

In the next few decades, these conifer-clad mounds became the most prominent reminders of the mining heritage of this rural area as all the old infrastructure of branch railway lines and pit heads were removed. They were moving from being sites of industrial waste to sites of industrial heritage *and* local ecology. Camerton batch was affectionately known as 'Little Switzerland' in the locality. As in the case of Victoria Square, development of the site involving the clearing of trees was considered in the era of 1980 s industrial restructuring. The National Coal Board considered proposals to clear the trees and recycle the spoil. However, local opposition was immediate and vociferous; the batch was a much loved and (unofficially) used green space – a landscape feature and marker of the lost mining days. Both the planted trees and those which had self-seeded onto the site from its margins became the foci of opposition to development on the grounds of conservation and heritage.

The parish council, which was already seeking to tidy up other derelict parts of the old mine site, ended up taking possession of the batch and adjacent land, making it into a nature and mining heritage trail. However the (closely packed) conifers made a very 'user-unfriendly' environment for such uses and so woodland grants were obtained to improve and manage the site. Given that this was the era when national forestry policy prioritised native broadleaf reafforestation, about a third of the conifers on the batch were clear felled. This caused further uproar in the local community and the management plans for total broadleaf regeneration were abandoned. So Camerton batch, which is now a local nature reserve, has a heritage and nature trail climbing up and down its steep slopes through areas of growing native broad leaf trees and areas of tall 110-year old conifers. Again, as in the square, place identity here seems settled for now. However, the precise form of that identity is less than stable, as many of the remaining conifers (which are very tall and thin) are reaching the end of their life and the original plan to convert the site to a more 'natural' native woodland may slowly re-emerge, even though this will diminish the site's character as one of the dark green batches which mark out the mining community heritage. It should also be recognised, however, that the cultural, political and emotional context in which the Camerton Batch is sited are also dynamic, such that the attachment to the authentic heritage of the site held by those with either personal or narrated living memory, may well also diminish among ensuing generations.

Fig. 5.2 Camerton Heritage Trail and Nature Reserve: then and now

The Cemetery: Arnos Vale cemetery was set out in 1837, on previously private parkland on the then edge of the city of Bristol. It was one of the new Victorian cemeteries which sprung up in the major cities of the UK as the old urban burial grounds failed to cope with growing populations. At first it was planted in the parkland style of Capability Brown but its dominant landscape form emerged in the middle decades of the 19th century, following the highly influential 'cemetery style' devised by J. C. Loudon. This specified a highly specific and ordered pattern of design and planting drawing upon a whole range of mostly evergreen trees chosen for their form, colour, shape and ancient associations. These were newly available to UK landscape designers as a consequence of the work of the Victorian plant hunters who had travelled the world seeking new tree and plant species.

During the latter half of the 19th century, the cemetery (and the newly planted trees) flourished, becoming the principal place of burial for the city and populated by spectacular buildings and grand monuments. It was commented upon for its order, dignity and solemn beauty.

However the 20th century marked a change in fortunes and became a story of slow decline and growing dereliction. The now 44 acre site was engulfed by the dense housing developments of the growing city. With some 90,000 internments, the cemetery represented a complex space requiring significant maintenance and management, especially when it became less popular due to the establishment of rival municipal cemeteries and the increasingly popular practice of cremation. Ownership changed hands twice, but the cemetery remained, unusually, in the private sector.

This period of decline was associated with significant changes to the site. Management gradually retreated from the older areas of the cemetery and the intensity of maintenance was severely reduced. Accordingly, large parts of the cemetery became wild and overgrown: the planted trees were often too large for their locations, enveloping the monuments they once stood respectfully alongside; and wild trees had colonised many areas, using the graves and joints in their masonry to get established. Community labour programmes helped cut back many of these trees in the early 1980s but this only served to coppice them and to clear away competing brambles, leaving a veritable forest of ash and sycamore saplings to spring up, slowly destroying many graves and monuments. The result of this chaotic landscape of trees was to enable the owners to suggest the potential for housing development on the grounds that the site no longer fulfilled its role as a well-landscaped burial ground.

However, a protest movement sprang up to defend the space, not only as a place of remembrance, but also to protect more recent uses of the site as an urban green space. Ecological surveys of the cemetery, including tree surveys, were used as part of the protest movement's ammunition against development and although the city council at first claimed itself to be powerless to intervene in the dispute, it was instrumental in designating Arnos Vale as a conservation area and its woodland as a significant landscape feature in the city.

Ultimately, the cemetery was closed by the owner (which in effect meant abandoning the site), but volunteers opened it up again, providing daily security at the gates and undertaking some repair work, although such activities were restricted legally. After much pressure and delay, the city council finally enforced purchase of the site with a Compulsory Purchase Order and the cemetery passed into the possession of a Trust which has already gained funding from the Lottery Heritage Fund for restorative work. Thus the cemetery has entered a new phase of identity in which many of the trees will be cut back to ensure the protection of and access to the more important and interesting monuments. Wilder areas will also be left, but the state of current dereliction means that changes to the site and to the tree population will take years to unfold.

Fig. 5.3 Arnos Vale Cemetery: then and now

Conclusion

In each of these three tree-places, we argue that the active materialities and capacities of the trees have been significantly influential, not only on how the sites have developed but also on human practices and performances within them. The ability of trees to grow, reproduce, spread, break up monuments, figure significantly in the emotions of nearby residents, demarcate heritage and so on have in each case slipped the leash of human plans. The trees have acted as relatively autonomous material presences which have spanned across and between eras of place identity and place configuration. In doing so, their powerful material presence has relationally shaped the new place identities and configurations that have emerged. New waves of politics, emotions, economics and governance have gathered around the trees and formed alliances (or otherwise) with them in disputes about future place form. The politics and ethics of these places need to deal both with local place-relations and with wider scales and distances of relations. For example, the quite necessary and in some instances unnecessary removal of trees from these places has needed to consider both the city and the countryside as a place of trees and the world as a place of trees. Trees deserve recognition because they help make the earth and cities and countryside, liveable, even if local politics might require that in the name of a just relationality, they be removed in some places, while in others, defended and loved with a passion.

The stories narrated in our three sketches reflect both specifically situated tree agency and also a wider sense that the material presence of trees with their active capacities will afford palpable contributions to how many places, and their shifting material forms, have unfolded over time. The richness of place defies capture (Buell, 1995). In the brief narratives of places in process which we have sketched out above, decades of history are jumped by in a flash. In those decades many people will have come and gone as internees, visitors, rememberers, mourners, workers and researchers. To them, at any given moment, the place seems solid and fixed. But over longer timeframes the places are clearly on the move. The lively material presences of the trees have acted as unruly threads working at their own speeds, bridging between land uses, connecting different eras of politics and economics, scrambling order into disorder and new order. The agency of trees and other non-humanity is not, of course, as reflexive as human agency, but it can be seen as creative; as meaningful. To account for the history and present condition of the above places without their active presences would be impossible.

The places presented (as all places) are not merely processes and narratives but whole *ecologies of interrelating trajectories* that settle into temporary local forms but which also have threads that weave through the local to global in scale. The ideas outlined above and the illustrative examples of trees in their places, allow us to argue that the agency of materiality and in this case the materiality of nature, needs to be factored into social explanations of the functioning of the world and its detailed space-time patterns. To ignore the agency of the non-human or merely to talk of social nature, is to help ensure that 'the world is rendered as an exclusively human achievement in which 'nature' is swallowed up in the hubris of social construction' (Whatmore, 2003: 165). Social construction alone is untenable in this respect, given that we are made of stars and live through microbes and amongst trees.

Acknowledgments The authors gratefully acknowledge the financial support of the Economic and Social Research Counsil for their research on "Arbori-cultures: the significance of trees to places"(R000237083)

References

Amin, A., and Thrift, N., 2002, *Cities: Reimagining the Urban*, Polity Press, Cambridge.

show Benton, T., 1989, Marx and Natural Limits, New Left Review 178: 51–86.

Braun, B., and Castree, N., editors, 1998, *Remaking Reality: Nature at the Millennium*. Routledge, London.

Buell, L., 1995, *The Environmental Imagination, Thoreau, Nature Writing and the Formation of American Culture*. Harvard University Press, Cambridge, Mass.

Callon, M., 1986, Some elements of a sociology of translation; domestication of the scallops and the fishermen of St Brieux Bay. In *Power, Action and Belief: A New Sociology of Knowledge?*, edited by J. Law, pp. 196–233. Routledge and Kegan Paul, London.

Callon, M., 1991, Techno-economic networks and irreversibility. In *A Sociology of Monsters?*, edited by J. Law, pp. 132–161. Routledge, London.

Callon, M., and Law, J., 1995, Agency and the Hybrid Collectif, *The South Atlantic Quarterly* 94(2): 481–507.

Cheney, J., 1994, Ecofeminism and the Reconstruction of Environmental Ethics. In *Ecological Feminism*, edited by K. J. Warren, pp. 158–78. Routledge, London.

Cloke, P., and Jones, O., 2001, Dwelling, place, and landscape: an orchard in Somerset, *Environment and Planning A* 33: 649–66.
Cloke, P., and Jones, O., 2004, Turning in the graveyard: trees and the hybrid geographies of dwelling, monitoring and resistance in a Bristol cemetery, *Cultural Geographies* 11: 313–341.
Demeritt, D., 1994, The nature of metaphors in geography and environmental history, *Progress in Human Geography* 18: 163–85
Fitzsimmons, M., 1989b, Reconstructing Nature, *Environment and Planning D, Society and Space* 7: 1–3.
Fitzsimmons, M., 1989a, The matter of nature, *Antipode* 21: 106–20.
Fitzsimmons, M., and Goodman, D., 1998, Incorporating Nature: Environmental Narratives and the Reproduction of Food. In *Remaking Reality. Nature at the Millennium*, edited by B. Braun and N. Castree, pp. 194–221. Routledge, London.
Game, A., 1991, *Undoing the Social: Towards a Deconstructive Sociology*. Open University Press, Buckingham.
Gordon, S., 1997, How many kinds of things are there in the world? The ontological status of societies. In *The Mark of the Social: Discovery or Invention?*, edited by J. Greenwood, pp. 81–103. Rowan and Littlefield, Lenham.
Haraway, D., 1988, Situated knowledges: the science question in feminism and the privilege of partial perspective, *Feminist Studies* 14(3): 575–99.
Haraway, D., 1989, A manifesto for cyborgs: science. Technology, and social feminism in the 1980 s. In *Feminism/Postmodernism*, edited by L. J. Nicholson, pp. 190–233. Routledge, London.
Haraway, D., 1991, *Simians, Cyborgs, and Women: the Reinvention of Nature*. Chapman & Hall, New York.
Haraway, D., 1994, A game of cat's cradle: science studies, feminist theory, cultural studies, *Configurations* 1: 59–71.
Harrison S., Pile, S., and Thrift, N., editors, 2004, *Patterned Ground: Entanglements of Nature and Culture*. Reaktion Books, London.
Harrison, C. M., and Burgess, J., 1994, Social constructions of nature: a case study of conflicts over the development of Rainham Marshes, *Transactions Institute of British Geographers NS*, 19: 291–310.
Harvey, D., 1996, *Justice, Nature, and the Geography of Difference*. Blackwell, Oxford.
Ingold, T., 1993, The Temporality of Landscape, *World Archaeology* 25: 152–74.
Ingold, T., 1995, Building, Dwelling, Living: how people and animals make themselves at home in the world. In *Shifting Contexts: Transformations in Anthropological Knowledge*, edited by M. Strathern, pp. 57–80. Routledge, London.
Ingold, T., 1997, Life beyond the edge of nature? Or, the mirage of society. In *The Mark of the Social*, edited by J. Greenwood, pp. 231–252. Rowman and Littlefield, London.
Ingold, T., editor, 1998, *What Is An Animal?* Unwin Hyman, London.
Ingold, T., 2000, *The Perception of the Environment. Essays in Livelihood, Dwelling and Skill*. Routledge, London.
Jones, O., and Cloke, P., 2002, *Tree Cultures: The Place of Trees, and Trees in Their Place*. Berg, Oxford.
Jones, O., 2006, Life in a Globalised World. In *of Trees and Trails: place in a globalised World*, edited by N. Clark, D. Massey, and P. Sarre, pp. 214–264. Open University, Mitton Keynes.
Latour, B., 1993, *We have Never Been Modern*. Harvester/Wheatsheaf, Hemel Hempstead.
Latour, B., 2004, Non-Humans. In *Patterned Ground: Entanglements of nature and culture*, edited by S. Harrison, S. Pile, and N. Thrift, pp. 224–27. Reaktion Books, London.
Law, J., 1994, *Organizing Modernity*. Blackwell, Oxford.
Macnaghten, P., and Urry, J., 1998, *Contested Natures*. Sage Publications, London.
Massey, D., and Thrift, N., 2003, The passion of place. In *A Century of British Geography*, edited by R. Johnston and M. Williams, pp. 275–299. Oxford University Press & British Academy.

Massey, D., 2005, *For Space*. Sage, London.
McPhail, R., and Ward, D. E., 1988, *Morality and Agency*. University of America Press, Lanham.
Murdoch, J., 1995, Actor-networks and the evolution of economic forms: combining description and explanation in theories of regulation, flexible specialisation, and networks, *Environment and Planning A* 27: 731–58.
Murdoch, J., 1997, Inhuman/nonhuman/human: actor-network theory and the potential for a non-dualistic and symmetrical perspective on nature and society, *Environment and Planning D: Society and Space* 15: 731–756.
Thrift, N., 1996, *Spatial Formations*. Sage, London.
Thrift, N., 1999, Steps to an ecology of place. In *Human Geography Today*, edited by D. Massey, J. Allen, and P. Sarre, pp. 295–352. Polity Press, Cambridge.
Tudge, C., 2005, *The Secret Life of Trees: How They Live and Why They Matter*.Allen Lane, London.
Warren, K. J., editor, 1994, *Ecological Feminism*. Routledge, London.
Whatmore, S., 1999, Rethinking the 'human' in human geography. In *Human Geography Today*, edited by D. Massey, J. Allen, and P. Sarre, pp. 22–41. Polity Press, Cambridge.
Whatmore, S., 2002, *Hybrid Geographies: Natures, Cultures, Spaces*. Sage, London.
Whatmore, S., 2003, Handbook of Cultural Geography. In *Culturenatures: Introduction: More than human geographies*, edited by K. Anderson, M. Domash, S. Pile, and N. Thrift, pp. 165–167. Sage, London.
Whatmore, S., and Boucher, S., 1992, Bargaining with nature: the discourse and practice of 'environmental planning gain', *Trans. Inst. Br. Geogr.* NS 18: 166–178.
Whatmore, S., and Thorne, L., 1998, Wild(er)ness: Reconfiguring the geographies of wildlife, *Transactions of the Institute of British Geographers*23/4: 435–454.
Wolch, J., and Emel, J., editors, 1998, *Animal Geographies*. Verso, London.
Woolgar, S., 1991, Configuring the User: the case of usability trials. In *A Sociology of Monsters?*, edited by J. Law, pp. 58–99. Routledge, London.

Chapter 6
Intelligent Artefacts at Home in the 21st Century

Richard Harper, Alex Taylor and Micheal Molloy

Introduction

For us, the term 'agency' is a label to define that property of an artefact that gives it a place in some train of human action. Now, from this view, an artefact might be given agency by dint of being hit with another object held by a person. Thus the mechanics of movement could be a label for an agency of sorts. But mechanics have no interest to us. Our view on agency assumes that the word is used to describe a sequence of acts that make manifest not just an action, a hitting of an object, but a thought, an idea, a moral purpose; in other words, a human action with a particular kind of intention.

Now this sense of the term agency is one that not only opens up all sorts of conceptions and possibilities, but also opens up all sorts of conceptual confusions. As to the first: conceptions and possibilities. It leads one to imagine that objects with certain kinds of agency could be 'tools of the mind'. This, in crude summary, is the credo of distributed cognition and its various acolytes. As to the second: confusions. It seems to us that this claim – viz.. how an object with agency is acting out an intentionality – has to be rigorously circumscribed. To say that a calculator is a tool of the mind, for example, is acceptable. But beyond this? The trouble is that in itself this claim seems too modest: after all, it tells us nothing that one would not ordinarily know insofar as one might pick up such a tool to help one solve a piece of arithmetic, say. Indeed one would expect a school kid of age 5 to know this. To say that somehow a device (say a calculator, again) 'extends the mind' leads one to want to say something more, though, and here confusion can slip in. Consider some of the things that one might want to say which claimed 'more'. One might want to say, for example, that an artefact, the calculator, has intelligence in itself. Another might be to say that an amalgam of things and a human mind are intelligent as a whole. The first of these statements few would aver to, one would imagine, even the most

---

R. Harper
Microsoft Research Cambridge, Cambridge, UK
e-mail: r.harper@microsoft.com

C. Knappett, L. Malafouris (eds.), *Material Agency*,
DOI: 10.1007/978-0-387-74711-8_6, © Springer Science+Business Media, LLC 2008

strident advocate of distributed cognition; the second most might accept at first glance, thinking it reasonable. But it is not, on our view. Let us explain.

In this view, intelligence is a quality or a quantity of the things in question. It is this (whatever this might be) that produces an outcome. Accordingly, this view might lead one to suggest that 'more things' in 'the right amalgam' might produce a 'righter', 'bigger', 'more intelligent' outcome. As it happens, this is more or less what Ed Hutchins claimed in his curious book, Cognition in the Wild (1995). There he argues that the ways a mix of artefacts gets used as an assembly to allow large boats to be navigated in tight harbours makes the captain of the boat (or more particularly the pilot of the same) intelligent; here 'intelligence' is like a calculation. The problem we have with this is that this is not how one uses the term intelligent. That an amalgam of persons and machines might allow the person to achieve something intelligently is a claim that is reasonable. But to claim that the things and the person make intelligence is, in our view, wrong.

This seems merely semantic, but it is consequential, we would argue. Consider: one might say a pilot who fails to get his or her boat in to harbour fails because one of his or her tools fails. But this is not the same as saying that the pilot is unintelligent. Nor is it the same as saying that the tools (any individual one or them as a whole) are unintelligent. All it is saying is that the pilot failed in their job for a known reason – a tool or a set of tools did not work. Of course, a situation where a tool fails might lead one to condemn that tool so that thereafter it is not used or, if it is, only circumspectly. Condemnation is important, it hardly needs any saying. But at no point would one be saying when one condemns a tool one was doing so because 'it' was unintelligent or even because it was part of an 'intelligent system'. Its condemnation has to do with the fact it did not do what it was (is) supposed to. By contrast, when would it be that one would say of a pilot that he or she was unintelligent? Not when the tools let them down, we have seen. But one would say so in the following sorts of situation: when a pilot failed to take notice of the fact that a tool was not working and did not devise work arounds or some other form of ameliorating action (like choosing not to go to that harbour). In this sense, a pilot's intelligence is held to account not for the actions of artefacts, whatever their agency, but because of his or her actions. In this view, intelligence is not a quality in a thing or a set of processes; nor is it a quality of agency. It is a question of moral culpability for persons. The term intelligence is, then, a moral term. Following Wittgenstein, we take the term to describe or label normative judgements: judgements about whether something has been done well, badly, thoughtfully or negligently. The term implies too that some one is accountable for the behaviour in question and for the arrangements of artefacts that allowed or enabled that behavior. Intelligence is thus never an abstract concept – neither a label for an abstract thing nor a term applicable to empirical objects other than human – it is a term only applicable for human doings (though of course, these doings often entail the use of artefacts).

Now, what has this got to do with agency? How did we get here, only a page or two from the outset of this chapter? We are discussing this now since we think that misappropriating what agency might mean and relatedly, the concept of intelligence, as bound up with the analytic purchase of the term agency, can lead to distraction and muddle in the ultimate tool we have at hand: namely language. Thus, when we say muddle and distraction, we do not mean that we say this as our view; we are not saying this as if it were our opinion; ours is simply a view amongst many. We are saying this in reference to how it is that when one stretches the limits and properties of language one can, sometimes, get wayward. Language is a tool of sorts but has to be used appropriately, we are saying; it is very easy, indeed, a chronic feature of words that they can be used in ways that misleads. And this is what we think is happening in the word agency gets used, most especially when it is used in connection with the word intelligence. When agency is used as a label to indicate something about the relationship between human 'doings' and the 'things that people use', it seems to us that muddles occur and what is meant, ordinarily, gets confused.

Does this matter? Well, not because one wants to honour language; but only because one wants traction and direction on one's own thoughts. There are a number of ways that one can lose such traction and direction. The most important one, at least the one we want to focus on in framing our chapter, is through making the workings of the mind (or at least perceptions as to what the workings of the mind might be), come to be far too important in the analysis of the relationship between things and people. We want to suggest that this is consequential for psychology (cognitive or otherwise), sociology, archaeology and design, to name but a few disciplines. It seems to us that when distracted by having the 'mind' as the focus, disciplines that ought to be concerned with other matters start looking at methods for circumscribing the functioning of the brain. For archaeologists, this means that the artefacts they retrieve become indicators of the physiological state of mind of ancient man for example; for contemporary cognitive scientists, human action in everyday contexts become a measure of how the brain processes information.

We need to make it clear that there is nothing wrong with looking for these matters in themselves: they are perfectly legitimate tasks. But it is a concern when this distracts from what might be more insightful concerns. At its worse, it can lead science to discern what can only be described as truisms: that electronic calculators are tools of the mind is one such claim, for example (though as it happens this is not one that has been made, as far as one knows, but it suffices to illustrate the species of claim we have in mind). At best it leads away from a concern with discerning what is pertinent to understanding the contexts in question. Thus, to get a perspicacious understanding of how ships' pilots do their job, one would do well to put aside concern with how the brain works and look instead at the social and cultural systems that have produced pilots, ships, harbours and navies. One should look at the institutional histories of these 'social institutions'; at the social processes, that is, the training of pilots and so on; in sum, one should only focus on doings and not mental operations. Of course, this is what makes Hutchins' book so curious: he says that one should look at doings to understand the mind at work and thus his book leads one away from cognitive sciences as practised before that book's publication. But what we are saying is that one ought to take that book more seriously than Hutchins does himself and abandon a concern with mental categories and cognitive machinery altogether. Do not worry about the mind at all, we are saying, just focus on the doings if you really want to get to grips with pilots.

Now, our concern in this chapter are not doings in the past or even in the present but as it were, doings in the future. We want to present our view on how to understand artefacts and doings, linked by the term agency, by presenting our view on a certain sort of technology design. In particular, we want to explain and account for a number of our own technologies that might go in to what is called, in contemporary parlance, a smart home and claim that these designs reflect an attempt to ensure that those artefacts let humans be smart, thoughtful and intelligent by making sure that accountability, culpability and the moral responsibility for action remain in the person's hands and are not merged, through conceptual confusion, with the 'agency of artefacts'. It is achieved too by eschewing all concerns with categories of the mind or with notions of distributed cognition and relatedly, how artefacts might be agents in this. Enabling human agency in the home of the future is our concern, certainly, but we want to show that how we get to our designs is through being strict about what we mean by agency and nested concepts like intelligence, in the design processes. Now, it has to be said that whatever our pride or vanity, the quality of our designs is not an issue here – they might be good or they might be bad. Our goal in this chapter is to try and expound the philosophical position that has led to their design so as to credit the relevance of that position. Others, better versed in the nomenclature of, say, ordinary language philosophy, might be able to better argue the critical view on the concept of agency and the concept of intelligence (e.g., Button et al., 2005); we have to argue with the devices we build. They will have to act as agents of sorts, though whether the reader will let them so behave is another matter.

Our Domain of Inquiry: An Archaeology of the Future

We choose the case of smart homes and devices within them for the very simple reason that interactive systems computing research has become, recently, very heavily weighted towards the idea of a smart home. There are now several ongoing, prolific and high-profile research programmes in this domain, including Georgia Tech's "Aware Home", MIT's "Place Lab", Samsung's "Smart Home Project" to name but a few. The goal of much of this research has been, to put it crudely, exploring ways of using computing to make homes more 'intelligent'. What has been meant by the use of these claims in the various academic papers that have been produced and in the marketing material that often gets associated with the projects is not always clear, though as we shall see, it is bound up with pragmatics. Leaving aside the technical merits of the ubiquitous and pervasive computing that has resulted from such projects, in practice the most obvious social or human benefits (still more hypothetical than real, alas) have been for those individuals who have constraints on their ability to effectively conduct themselves intelligently. We are thinking here, most expressly though not unkindly, of the handicapped, the aged, the sick and so on. With the smart homes that have been built, disbursement of medicine can be monitored and managed; accidents that might befall old people can be observed and medical intervention summoned if necessary. In this way, what was before merely a set of walls and enclosed spaces has become, in these smart homes, an infrastructure with what some have called a 'technological intelligence': the ability, of the machines therein, to monitor, look and act at appropriate moments.

Leaving aside, for the moment, what one might think of this particular view of 'intelligence' and its agents, achieving these benefits requires a specific type of networked-smart home experience, replete with sensors, monitors and cameras of various kinds. Unfortunately, for home dwellers without the particular needs of the aged or sick, both the complexity of these technologies combined with their unfamiliarity demands a very high level of perceived benefit before they become appealing. Indeed, if the history of research into this area attests to anything, it is the narrowness of the appeal of smart homes to a wider population (Abowd, et al., 2003; Aldrich, 2003; Edwards & Grinter, 2001).

Nevertheless, this is not of concern to us here. As we say, we have a different view on what smart homes might be which does not pertain to the question of which segment of society they are to be built for. For us the goal of 'smart home research' is not to design technologies for specific and unusual needs through the implementation of networked technical infrastructures. Rather, we start from an altogether different assumption: we think that the home is already and always has been a place where people can be smart, smart not in terms of technology, but in terms of how people conduct their lives. Our view is that the smartness or, if you prefer, the intelligence in the home is to be found not in the artefacts nor in the structures that the artefacts are part of; it is to be found in the human endeavours with those artefacts. Recognising this, our approach to the invention and design of devices for the home is to design them so as to augment and support the uniquely human art of home making. The intelligence we seek to enable is not in the devices then, it is in the artful ways that those devices are leveraged by the occupants. This way of thinking, as we will show, results in technological concepts which are quite different from those typically found in smart homes which often seem to be based on blurrings of the term intelligence with agency; a blurring of the moral culpability of human action with the concept of mental functioning supported by embodied artefacts. Hence our agitation at the outset for trying to conceptually define what agency might mean. In our world, our research world in computer human interaction, agency and the smart home often mean intelligence in the artefact, agency as a synonym for distributed cognition (for a parallel argument, see Crabtree, et al., 2003; Tolmie, et al., 2002).

We wish to present our own view not by reporting all the designs we have worked on but by exploring a particular theme that highlights how we see human intelligence at play. More especially, we focus on what might seem to be the most obviously unintelligent and mundane aspect of homes: namely surfaces such as fridge doors, notice boards, kitchen walls and even the surfaces provided by bowls. We explore how these surfaces are transformed from being merely the materials that constitute a house into resources for the organisation and enrichment of home life but without having to impose some categorizations of agency and intelligence. Having understood this, we then elaborate on the ways in which we have extended the power of these surfaces as instruments, through the use of computing, again without any conceptual baggage.

Substance in Design

We are, the reader might be surprised to learn, not the first to focus on such prosaic properties of home settings as fridge doors, walls, surfaces and bowls. Nor are we the first to explore the potential of digitally augmenting them. Attention has been given to picture frames, for example (Kim et al. 2004), digital pin-boards (Laerhoven, et al. 2003), and much else beside (e.g., Bonanni, et al. 2005; Ju, et al., 2001; Petersen, et al., 2005). What makes our research different though, and this does relate to our concern with what intelligence might be, is how the use of any particular artefact is subject to a larger regime of behaviours: in our view, to understand where a picture frame is located, for example, is to understand the moral order of the home where the location of a picture tells one something about how the occupants of that home want to communicate their intelligence. They place some pictures 'here' because they are thus on show to visitors for instance, they place other pictures 'there' because they are in a more private place. Function, purpose, role, use, are bound up with the ecology of place, then, which is as much a physical system as a social and moral one. This harkens back to Norman in the *Psychology of Everyday Things,* who noted how placement of information in particular places can act as a memory aide (1988) but adds to that a sensitivity to the delicate arrangements of meaning that direct, control, guide and provide a framework for understanding the intelligence behind the physical. And here what we mean by intelligence is how the artefacts in question owe their existence, location and function to what someone has judged to be 'right for the home'. In each of the following examples of technology concepts we shall explain how we have sought to design devices that allow people to make these decisions as *they* see fit, so as to allow *them* to act intelligently.

Fridge Surfaces and Augmented Refrigerator Magnets

The first example we would like to turn to emerges from our field studies of fridge surfaces in family homes. As most readers will know from their own experience, one of the notable properties of fridges is their relatively large display surfaces (although this can and does vary between houses and indeed between countries). Indeed, if fridges are oriented in certain ways, households can find they have two and sometimes three large surfaces available: the front and whichever sides are accessible. This expanse of space can be put to good effect in some simple but nevertheless useful and in our view intelligent ways.

For example, a fridge's different physical regions can be assigned to particular uses and even allocated to particular people. Hence, the lower regions of family fridges are often taken over by items belonging to children, while higher up, it is common to find 'working' areas containing shopping and to-do lists and other organising items for 'Mum'. Scattered across these zones, more often than not, are memorabilia (Fig. 6.1).

In some cases, the divisions between the different regions can be more formal. One side of the fridge might be given to a household's organisational items, the front to family photos and the lower areas to children's things (Fig. 6.2). Such

Fig. 6.1 Haphazard display of photos, artwork and invitations

Fig. 6.2 Working area to left plus "family history" displayed on fridge door

organisation has the advantage of making it clear whether items are associated with specific activities or belong to particular household members. Spatial patterning can also be used to signify the change in status of items: a party invitation moved from a fridge's working area to its family display area can signify that the action has been taken to accept the invitation, for instance. Regardless of the particular arrangements, the salient point is that fridge surfaces lend themselves to having an array of heterogeneous items attached. The fridge's form—the height of its surfaces and its separated sides—helps in offering a simple way to categorise materials. All of these arrangements, enabled by the fridge's form, are controlled by those in a house.

Of course, it could be that any surface in the home would afford the same utility; but part of our approach is understanding what about the particular site of a surface gives it properties specifically relevant for the human endeavour in point. As we have mentioned, our own research and others have noted that where a display surface is situated in the home is key to understanding what is displayed, as well as when, how and to what ends. The same holds true for fridges. In nearly all homes, fridges are in the kitchen and thus seeing fridge surfaces is an almost unavoidable consequence of ordinary life—when preparing meals, making drinks, snacking and so on. Indeed, in most homes the established moral order, if you will, gives household members the right to be in the kitchen, use the fridge and to consequently view the contents attached to the fridge, whether they want to or not. The fridge provides a surface which is not only 'public', but also inexorably interleaved with the rhythms of the home.

What we see then is that fridge doors and sides become interactive surfaces of a particular sort, holding some materials, but not others; affording a particular range of interactions that weave into ordinary routines. In short, the physical form of fridges and the way in which use of that form is embedded into a home's social organisation set it apart from other surfaces. Surfaces on fridges become, in our terms, surfaces of a special kind not in what they do, but in the way they are used. Our claim is not merely that, in the home, various surfaces are 'interaction and display points'—this much is obvious. This discussion of fridge doors and sides is intended to show that what makes homes what they are is how surfaces (amongst other things) are used to display material in particular ways. Intelligence, if one wants look for it, is in the human art of deciding where things are put, how those things are put, and with what intended effect. Fridges may be dumb, but the way artefacts are attached to them is not.

With this in mind, we have considered ways that we might assist these intelligent doings with digital means. Rather than substituting fridge doors with digital alternatives, we have been exploring ways of letting fridge doors (and sides) let the user do more. We have begun by focusing on a distinctive property of fridges that enabled the intersection of where, how, what and when: namely that most fridge doors are magnetic. Magnets allow all manner of items to be attached to fridges; things can be attached anywhere with little to no thought and their movement and removal is made trivial. This enables a fluidity to fridges as a display—things can be easily moved into, around and between the different regions and no prescribed arrangement is enforced. The design concepts we have derived from this seek to augment this magnetic property to further enhance the fridge surface's useful functions.

In the first instance, we have conceived of reminding magnets. Somewhat perversely, sometimes things left as reminders on fridge doors get forgotten about. Although fridges are good places to put reminders, they are not necessarily perfect places. Our reminding magnets are a lightweight solution for drawing attention to items in unobtrusive ways. In one version of the concept, moving the magnet causes it to glow for some period of time thereafter, drawing people's attention to items that are newly attached or newly rearranged (Fig. 6.3a). In another version (Fig. 6.3b), magnets that glow on specified days can be attached to items which need to draw attention to themselves on those days such as appointments and party invitations.

A second concept builds upon the practice of putting important and frequently used items like shopping lists and school term dates in specific places on fridge doors. The fridge-glance concept is designed to overcome the problem of accessing this material when away from home; for example, when shopping or making calendar arrangements. Incorporating an in-built camera, the concept allows items placed within a purpose-built magnetic frame to be remotely viewed via a camera-phone or Internet browser. The frame is meant as a visual cue, a mnemonic, demarcating an area where items can be casually placed to be remotely accessed. In this way, the design takes into account the informal, offhand use of fridge surfaces, but remains sensitive to the importance of particular attached materials.

A third concept, talking magnets, is intended to help 'annotate' the materials placed on fridges. Annotations might be useful when additional information about an item might need to be quickly and easily conveyed, to dynamically label a magnet (e.g., changing the status of items affixed to a fridge,). Differences in design might reflect differences in purpose: one might make it clear who a message is from (Fig. 6.4a) and another merely provides the message itself where identity might not matter (Fig. 6.4b).

At the moment, these are initial concepts, attempts to explore how we might augment what people do with particular surfaces, though even as we write prototypes are on display in our 'lab'. As is the way with technology, however,

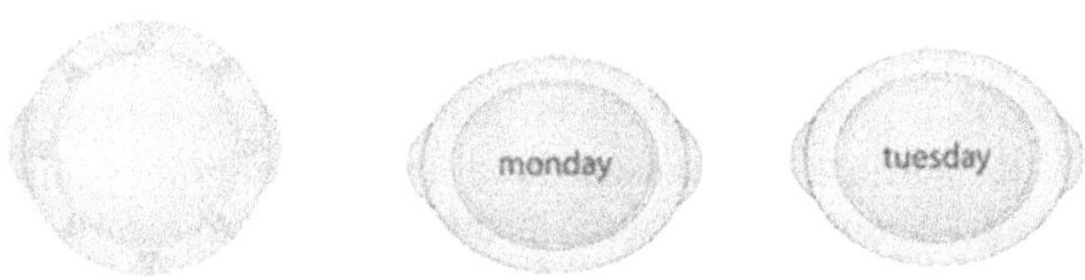

(a) (b)

Fig. 6.3 Magnets that glow when moved from one place on the fridge to another, drawing attention to attached item(s): (**a**) glows for 24 hours once moved, (**b**) glow on the labelled days

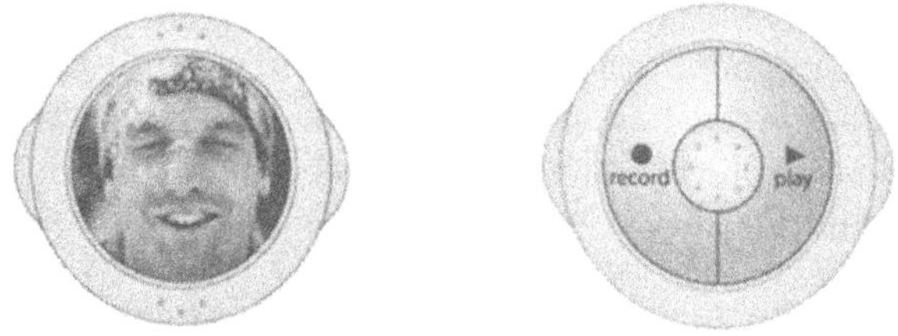

(a) (b)

Fig. 6.4 Magnets that allow items on fridge surfaces to be annotated: (**a**) indicates who created it and (**b**) allows voice recordings

building ones that do not break or look too cumbersome to make appealing is not always easy to do. Nevertheless, our ideas consist of using digital means to give greater conspicuousness to reminders, making remote access possible and allowing 'digital' annotation. We are not, at this stage, certain that these are all or even the best ways one might achieve the enhancements we have in mind. But in initial paper prototyping exercises, the overall response to these concepts has been largely positive, with five different households interviewed having their own particular favorites. Of relevance to the general theme of surface ecologies has been the reaction to the straightforwardness of the designs. Households were struck by the inherent simplicity of what they initially thought to be yet more 'technology' for the home. They responded positively to the idea that they might be able to operate the magnets almost without thought. This provoked a sense that the magnets would compliment the ways a fridge, as a surface, is used. As the mother in one household put it: "The most important thing is that they 're easy to do, that you don't have to turn them on. You can use them on your way to the sink to do the dishes or something."

Situated Messaging in The Home: Homenote

Turning from refrigerator surfaces, our second example relates to another kind of messaging, though this time going beyond the boundaries of the paraphernalia found on fridges. We are thinking here of how certain "low tech" artefacts, such as paper notes and Post-It notes, whiteboards, corkboards and paper calendars are used for within-home communication, for messages between members of a house when they are at home. These particular forms of messaging, some of which appear on fridge doors but elsewhere as well, are strikingly non-computational. These mundane artefacts also have distinct properties. As we have already described with refrigerator surfaces, the placement of these artefacts within the home, both physically and socially, is critical to their use. A note placed on the refrigerator door (and even where on the door it appears) has implications for who will see it and how it will be used. Further, people make particular decisions about where best to leave a note for someone else, there often being places in the home or "communication centres" where important messages are left (Crabtree, et al., 2003; Harper & Shatwell, 2003). Additionally, it is in the nature of these artefacts that, because they are inscriptions on paper, or on other display surfaces such as a whiteboard, they have a visual, static persistence or "epigraphic" quality to them. There are two implications of this. First, they attract attention to themselves in the periphery and as a consequence of everyday activity. For example, notes are placed in such a way that the right people will "come across" them when they are needed and in the course of their routine activities. Thus, they are "pushed" to people's attention in often subtle ways. Second, because they are visually displayed in this way, depending on where such notes are placed, they can be accessible to anyone present in a particular room or area of the house. Thus they are, in a sense, publicly "broadcasting" to no one in particular, but to anyone present.

Contrasting this with remote communication technologies, we can see at afford once that many of these "placeless" rather than situated messaging. The mobile phone and email, in particular, are "person-to-person" rather than "person-to-place" messaging. In other words, if I send an email, I have no real assurance where my message will be received, only who will see it: it may be that it will be read at work, at home or even on the road. If I call someone on their mobile phone, I have no real certainty about where and under what circumstances that call will be received. Remote communication technologies also differ in that they rarely push themselves to attention as a backdrop to other activities. They are more often foregrounded activities demanding attention, caused by the ring of the telephone, for example. A final difference is that remote messaging is often dynamic, transient and hidden from view (such as voice messages or email), and so sometimes does not naturally lend itself to broadcasting to more than one person or indeed to a household.

Our development of a prototype technology called HomeNote was motivated by the proposition that the unique affordances provided by paper-based messaging in the home, combined with the ability to remotely create them, would generate some compelling new design possibilities. More specifically, in building HomeNote, we wanted to explore the unique affordances and potential value of person-to-place as against person-to-person messaging technologies in the home. But we also wanted to deploy HomeNote into real households as a kind of "Trojan horse" to allow us to deepen our understanding of home communication. This in turn we hoped would allow us to explore possibilities for new and different concepts based on our understanding of the communication regimes of households.

As a starting point, we based HomeNote on TxtBoard, a situated messaging device that used the SMS protocol to let members of households broadcast messages home (O'Hara, et al., 2005). This device was expressly designed for simplicity with many paper-like functions. An early trial of TxtBoard with one household provoked some of the kinds of home communication we have discussed. We thus sought to combine TxtBoard's minimal set of functions with properties that might leverage new benefits. Specifically, given that so many of the messaging tools in the home involve inscribing in one way or another, we wanted to build a prototype that supported stylus markings or scribble, in addition to SMS.

HomeNote itself was constructed from off-the-shelf technology: it was a tablet computer encased in a wall-mountable frame containing GPRS and SIM cards. This provided each HomeNote device with a unique phone number to receive and display text messages from mobile phones. Because it was a tablet computer, the devices also supported locally scribbled notes or scribble annotations on top of text messages. Users could also switch between messages using the tabs along the top of the screen, create and delete new messages and see at a glance who sent a text message from the information down the left side of the panel (Fig. 6.5. (See Sellen et al., 2006 for more detail). In total, we built five prototype devices and deployed them in local households for a period of a month or more.

We found that HomeNote did extend the ways messaging practices were undertaken and not simply by combining remote delivery with local display in ways that prior surface technologies like Post-It notes could not. It also encouraged new forms of messaging. That is to say, HomeNote did not just stretch the intelligent use of particular kinds of surface, it helped create new ones.

For example, in supported remotely-created situated messaging, HomeNote demonstrated value for all households using it, allowing them to communicate in new ways. Thus, teenage children could send messages home to reassure all the family of their whereabouts and husbands and wives could text home to say that they were on the train and due home at a particular time. These kinds of messages were not only functional but they were also ways of having a presence in the family and expressing affection for other family members.

Aside from messages of awareness and reassurance, we also found many messages were, in effect, "calls for action" sent remotely to the household. Here

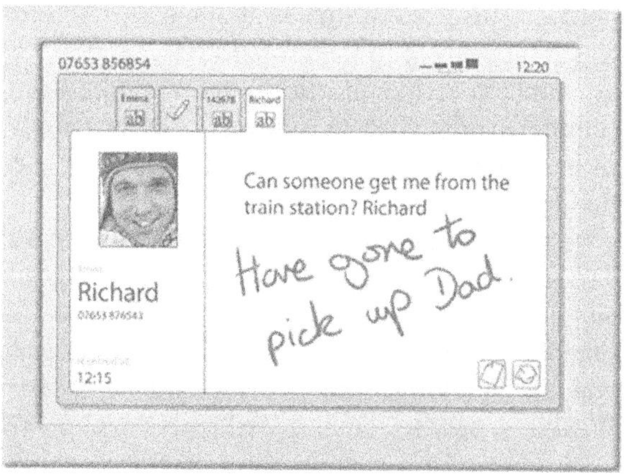

Fig. 6.5 The Home Note interface showing a text message overlaid with a scribbled note

we saw that the ability to remotely create place-based messages in the home was also used to valuable effect. HomeNote allowed calls for action to be finessed in new ways. For instance, in one household, one of the daughters would use HomeNote to request a lift home from her shift at the hospital. Here the fact that this message was posted in the background of ongoing domestic activity, broadcasting to but not specifying either parent, meant that such requests were viewed as less demanding than might have been done via the telephone. According to this daughter and indeed her family, the peripheral awareness afforded by HomeNote messages enabled an expressly polite kind of request to be made.

HomeNote was used frequently to broadcast what we came to call 'social touch' messages to the family. These were "I'm thinking of you" notes sent generally to a whole household or addressed to one person in recognition of the fact that others would see it. These would sometimes take the form of scribbled notes but other times would be sent remotely as text messages. For example, the father in one household regularly sent messages late at night to HomeNote to say "good morning" to his whole family, who could see the message the following morning or from work to say "welcome home" when he could not be there in person. Thus, the creation of notes remotely, with many of the paper-like qualities we have described, offered a new set of affordances for households.

In addition to demonstrating the value of remote messaging, we also found that, because HomeNote supported local scribblings created in the kitchen, it took on the role of a whiteboard, being the place for jotting down reminders, important telephone numbers, shopping lists, phone messages and so on. More interesting, however, was how HomeNote highlighted new kinds of messaging not previously recognised in the literature on communication (Sarvas, et al., 2005). For example, we found frequent use of the device for messages that looked like social touch messages but in fact were more about broadcasting the identity of the creator of the message, rather than directed at anyone else. Many of the 'good morning' messages scribbled by younger members of households, for instance, were signed with a flourish—as if these declarations and ornate signatures were intended to put that person's 'stamp' in the kitchen. Scribbles in which children announced they were off to bed or had finished exams were also of this nature, drawing attention to themselves without any particular purpose. We found that it was not just children seeking a visible space for their expression; the father who regularly sent 'good mornings' to his children complained when his messages were occluded under others or were scrawled on by children. In short, we came to the conclusion that these kinds of messages were playful, sometimes tender ways, of seeking affection or of drawing attention to their creator. They were, if you like, a form of saying 'don't forget me'.

In summary, this (relatively simple) prototype and its deployment underlined the ways in which the kind of communication that goes on in families is bound to place: to the sensitive—intelligent in our language—selection of particular places to put messages. But in addition, the introduction of the device encouraged a sensitivity to new forms of communication, where "placeness" was linked to affect. Family members appreciated it when others expressed a thought for them. They appreciated it all the more when these thoughts were displayed in a public place: the kitchen. Similarly people felt a tenderness (even a sympathy!) for those who simply messaged, in effect, 'think of me'. In other words, what we enabled with Homenote was intelligence of a kind, albeit sometimes sentimental. Smart homes should be as sensitive to this as they are to reminding, planning and other more functional types of tasks. After all, intelligence is not merely a matter of practicality; it is also a matter of affection.

Supporting Family Awareness: The Whereabouts Clock

Our next example leads directly on from our work around situated messaging in the home. Our studies of HomeNote made us realise that there was a place that could allow a variety of forms of expression, ranging from the tender through to the functional. But amongst these messages it became clear that some, a particular kind, had a special value that needed protection for themselves. This was not because these messages were notably valuable or rich, rather that their value derived from being seen at a glance in the place in which they were relevant.

In particular, the messages in question were related to the whereabouts of the sender. Here it turns out that where someone is not something that lets them express something in particular, it is a statement of fact that in itself expresses meaning. Thus the fact that someone is stuck on a train might mean that the person in question is lamenting the chaos of the traffic system, but what it also means (and this is more salient to those in the home), is that that person is likely to be late. This is of importance to the recipient of this information, important given where they are. It might mean that they might adjust their plans for eating, for example, or it might mean that they avoid worrying about the due arrival of their spouse or offspring. The whereabouts of people turns out to be a useful piece of information in the home for those doing particular things in the home.

It is in light of this that we built a device called the Whereabouts Clock. This allows family members to observe the whereabouts of other relatives, using a coarse-grained representation. Figure 6.6 (left) shows a screenshot of this initial prototype with three broad location categories identified as "Home", "Work" and "School". The middle of the circle identifies when individuals are travelling between locations or are in an unregistered location. Icons identify individual family members, these moving between categories to reflect changes in physical location. Our current implementation uses the identification of nearby cell towers to detect, at a loose approximation, where individuals are. Groupings of cell towers are mapped onto these high-level human interpretable categories. A SmartPhone client (Fig. 6.6 top right) scans for cell towers in proximity and sends updates via SMS to the situated surface when people cross over these mapped boundaries.

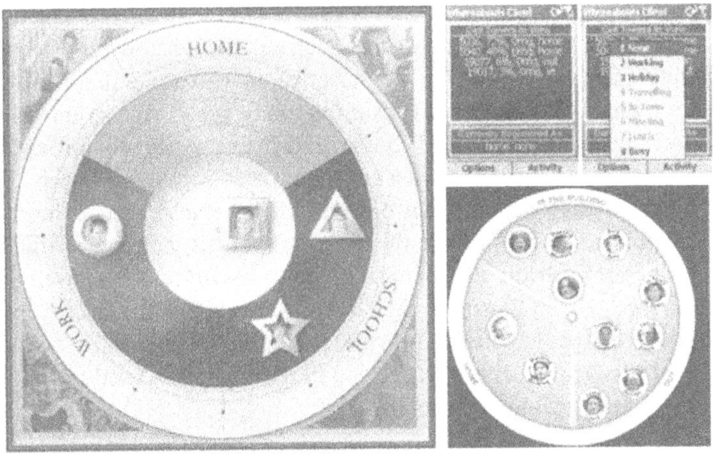

Fig. 6.6 One particular design for the home version of the Whereabouts Clock (*left*), office version (*bottom right*) and SmartPhone client (*top right*)

We designed the Whereabouts Clock to offer the protection for whereabouts information we had in mind, a protection which reflected its rather particular importance. First, and perhaps most obviously, the surface is intended to be situated in the home rather than remotely accessible or mobile. Though this is obvious, it is worth noting because it makes it clear that the information that this surface displays is designed for recipients in a particular place. The display is also intended to be always on, continually available for people in the home (and specifically in the kitchen) to view. This is not because we think this information is so important that it is needed twenty-four hours a day, but because its persistent availability means that it is there, when it is needed, whenever it is needed. A surface such as the one on the Whereabouts Clock offers visual information persistently and does so through being at-a-glance. Hence people at home can engage with it in much the same way as they might glance at a table or fridge to see if mail needs to be attended to. But, in being separate from these other places where various types of information might be located, the utility of this information is made greater. In this case, seeing is the trick; seeing without effort. Location information is important, but only to the extent that it can be seen without effort. The Whereabouts Clock reflects this.

We have as yet to trial the Whereabouts Clock either in the office or in the home. Whatever its fate, the fact that it is a situated display may be important in addressing some of the concerns around privacy found in the literature on location tracking. With our device, only people located in a particular place can view location information. These people are in the home and are therefore subject to its constraints: determining who can see (by dint of access rights to the house), as well as when they might be able to see (by dint of when people ought to be in the house) and so on. The result of this is that, by design, sensitive information is only broadcast to other trusted family members. This is not to say there are no privacy concerns around such a surface. There may be instances when family members prefer to be selective in revealing their location or activities to others. Teenagers, for example, may prefer not to have location information automatically pulled, but would rather push this information to their parents at select times to reassure them or they may be happy for their parents to know they are in town but would rather not reveal their specific location in a bar or club. We have therefore deliberately selected a high level of granularity of the location information in question. In this sense, we have tried to attain a level of 'intelligence' that is appropriate for the home, given the nature of the need for the information and the type of people who have access to it. Intelligence is as much about where you find out about something as much as what it is you know wherever you are.

Media Containment: The Picture Bowl

The final example from our studies is of a seemingly persistent feature of family households which would not seem to have much to do with surfaces. After having our attention drawn to a series of bowls holding a collection of miscellany in a household we were studying, we became intrigued by the different ways people collect, store and manage clutter. The prevalence of clutter in family homes seemed to recommend it for study if only because of its near ubiquity. As we began to delve into this topic, we also began to see how surfaces of a kind might be leveraged to offer new ways of dealing with clutter. To get to this point in our thinking, though, requires us to take a look at what clutter might be.

Clutter is made up of a variety of things: things temporarily out of place, things with limited life spans, things with ambiguous sentimental value, things in transition and things that no one knows what to do with, to name a few. By its nature, clutter in family homes is particularly heterogeneous because it represents the detritus of all the various family members. In and amongst a family's clutter, one finds functional things like glue, rubber bands, tape, lumped with children's broken toys, old sentimental items that do not quite deserve a place on the mantle piece and so forth. Similarly, there are coupons, batteries and cheque books sitting alongside what might be seen as the quintessential item of clutter, keys that belong to no obvious keyhole, but no-one dares throw out.

People deal with clutter in a variety of ways. They enlist bowls and drawers, dividers, tubs, plastic bags and all sorts of categorisation methods (or hardly any at all). How people in families choose to divide and store their clutter varies, as does the amount of effort expended, but what remains consistent is the use of artefacts that physically contain. The trouble with clutter, as we all know, is that it can spread out; bowls, drawers and the like keep it together, contained.

Although clutter is often treated in an off-hand way, what is evident is that where containers like bowls and drawers are situated in the home does matter.

And this returns us to our concern with surfaces, to the idea that where things get put, what things, when and how, is a measure of the human intelligence in a home. Here, though, this intelligence relates to being tidy, being, as it were, organised sufficiently that the home does not submerge under chaos. This is a kind of intelligence which is also (on the other hand) not so organising that it becomes a burden. One can be intelligently lazy after all.

It is possible to imagine a smart home automatically sorting and dealing with clutter in the ways we have described. The premise might be that a task that requires just a little bit of forethought and intelligence could be done away with and given, so to speak, to the building. From what we have seen it seems that such a solution would encounter all sorts of problems. It is evident, for instance, that the allocating of certain sorts of stuff to clutter bowls, drawers and so forth, is a thoughtful activity, where subtle judgements are made about stuff that may have no immediate place or certain 'home'. To be sure, some of it might eventually be given somewhere to go, but a lot of clutter sits in the bowls or drawers waiting until time and a little sentiment move it along, perhaps to another bowl or a drawer or sometimes to the rubbish. Thus, however intelligent a smart home might be, it is in the very nature of clutter that a proportion of it cannot be sorted out, that it remains ambiguous.

If this seems reasonable, it still remains some way from the design of technology. 'Some things don't have places to go: so what!', one can hear the smart home designer says so. But this is to miss the point that an intelligent way of dealing with the uncategorisable is required in the home. Stuff like keys for unknown locks is one thing, but it seems to us that at a time when members of homes increasingly carry all sorts of digital devices, the amount of digital clutter they bring home is increasing, too. We think smart home designers might ignore this clutter at their peril and though they might prefer to ban it from their smart homes, a solution for dealing with what one might call digital clutter is required.

Currently, the established solution for handling the proliferation of digital media (e.g., digital photos, video, music, etc.) centres on the PC. The PC serves as a 'hub' to peripheral devices designed to capture and play digital media, devices such as still and video cameras, MP3 players, PDAs and increasingly, mobile phones. There is, undoubtedly, much to recommend the PC as a destination for digital media. It offers a common interface to store, organise and manipulate digital media and gives users the ability to perform a number of sophisticated editing procedures. Seen from our perspective on the use of bowls and drawers, the PC, however, does not present an easy, low effort method for dealing with digital media. Rather, it offers what one might say is too much, an unwieldy piece of intelligence that does not reflect the casual storage and loose organisation that clutter deserves—even of the digital kind.

Take, for instance, the burgeoning use of media-enabled mobile phones. The content on these phones is not necessarily captured, stored, shared and occasionally cherished for its quality or to use in later editing. Instead, the quickly snapped photos or shared video are retained, temporarily or possibly for longer, primarily as a way of augmenting the lived experience of any moment in time

(Harper 2003;). Accordingly dozens of images are taken during a day, most of which have no value after they are shown. Some though, for a variety of reasons, may have value but this might not be clear at first nor something that the person who has taken the images wants to decide upon there and then. Instead, a common practice with mobile phone content is for users to keep the images on their device until they are forced to make a decision. Unfortunately, they are eventually confronted with the PC option, downloaded them all and edited. Our point is that this is a step too far. And here is the rub: what would a reasoned alternative to the PC be? It seems to us that what is required, instead, is a solution that reflects how images on phones have clutter-like properties. They consist of a mixture of stuff, some that has no value and some that does. What is needed is a way of putting this stuff somewhere temporary. This is what we have being trying to devise.

Drawing on the ideas we have developed from our study of clutter, we have attempted to determine what physical properties enable the low-effort storage of clutter. By enabling comparisons, our position on surface ecologies has been instructive. For one, it has provoked the question as to why it is that bowls and drawers, as opposed to other places including the innumerable surfaces one finds in the home, afford the practices we have described above. Why are countertops, floors or stair landings not the sites for clutter? The answer is obvious, but we will restate it: stuff needs containing.

Second, the idea that there is an ecology of surfaces in the home made us recognise that where a bowl of clutter is affords something special: it is placed in a way which reflects where the clutter may be 'properly unpacked'. We find that bowls in entranceways to the home, for example, succeed in their rolls as containers for keys, cheque books and the like because that is where those things spew forth from pockets and raincoats. Again, we return to the moral order of the home, to when and where it is acceptable to do things like place and amass clutter. Bowls and drawers, placed in particular locations offer just enough to deal with clutter where it arises and shows itself. Situated as they are, in the right place and at the right time, bowls (and indeed other containing devices) allow for an intelligently low-effort method of maintaining order.

Our studies also made us reflect on the fact that bowls display at least some of their content. This means that what class of things might be placed into them is visible to everyone in a house. So, whilst bowls contain, they also reveal; passers-by, as it were, can see what they are for and leverage it for themselves. In these ways, the placement of bowls, the way they display clutter, makes the organisation of tidiness tractable. It may be a form of idleness that leads people to throw things in to a bowl, but it is an intelligent way of dealing with the problem (of clutter) in home settings.

With these points in mind we have designed the Picture Bowl (Fig. 6.7), an augmented bowl that exploits how bowls 'work' and that further enables simple and lightweight actions for viewing and holding digital media. Still at its concept stage, we plan for our Picture Bowl to allow physical and electronic devices to be placed in it and when this happens for their content to be displayed

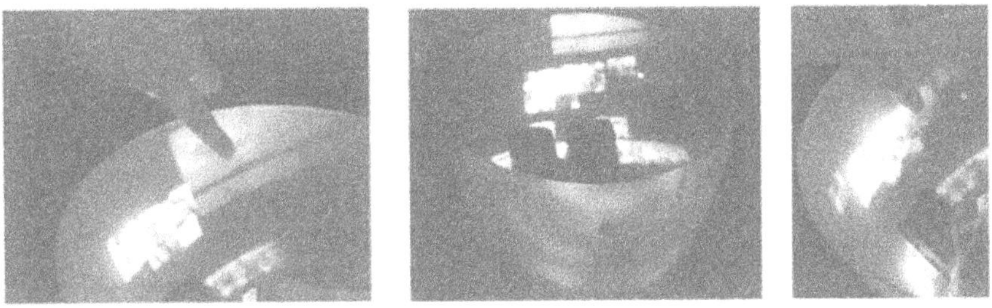

Fig. 6.7 Current manifestation of *Picture Bowl*. Two data projectors project media thumbnails onto the opaque glass surface where media is visualised. We anticipate using Bluetooth and to identify individual devices and transfer content. We also intend to touch-enable the bowl, incorporating a flexible transparent capacitive overlay

in the form of thumbnails on the sides of the bowl. As more devices are added, existing content will be 'pushed' towards the bottom of the bowl. In this way, the Picture Bowl will provide a sense of sidedness and depth—in essence a place to contain. We also propose that content will be copied to the bowl by simultaneously holding a collection of thumbnails with one finger and removing the associated device, simulating a peeling-off like effect. We have in mind how this operation will offer a low effort solution to simply shedding content, for instance in an entranceway bowl as one rushes out the door with a digital camera. The possibility of this stands in stark contrast to the efforts needed to upload content to a PC and, in doing so, being immediately directed into an 'environment' where one must organise and edit media.

To support the 'glanceability' of content in bowls, we also envisage thumbnails being slid up the Picture Bowl's sides and 'attached' to its top edge by selecting on and moving a thumbnail with a finger. This could allow specific media to be left for passers-by to see, possibly offering a subtle, visual reminder for some action or event. To further support this, we want to exaggerate the stretching effect that results in moving a thumbnail from the tightly curved bottom of the bowl to its relatively flat sides: as a thumbnail is slid up the side, we propose to increase its size and also its brightness.

Last, but not least, we imagine the bowl being portable so that it can be situated in places that fit the home in question. Ideally, a home might also have multiple augmented containers that could be situated to support different uses. This would allow, for instance, problems of privacy to be dealt within a common sense fashion. People could place personal containers along with their content in private places like the bedroom and thus privacy would be managed through the social ordering of the home and not through the cumbersome and arcane use of passwords and access rights. A portable device would also allow media to be moved from one place to another. Thus content might be brought to an augmented tabletop where it could be 'poured' onto the larger flat surface and organised, shared or deleted. This would further harness the properties of different surfaces, making the most of bowls for containing and the flat horizontal surfaces for activities such as sharing and organising.

Conclusion

And thus, in a roundabout way, we have come back to the beginning, to the idea of what intelligence might be as a concept driving a design orientation. Things do not have intelligence, in our view; things never will, we believe, because it is what people do with things that endows the human user of the things in question the ability to display their intelligence. For that only they, the human, the 'user' in our parlance, are culpable. Artefacts never can be morally responsible.

We have illustrated our arguments with concerns that have been deliberately mundane. We have wanted to avoid any muddles about complexity and function and form so as to focus on the conceptual issues at hand. Thus we have looked at 'surfaces in the home', for example, because they are places in which the 'intelligence of people' in the home is marshalled, displayed, leveraged and worked on. People use surfaces in intelligent ways to do various sorts of things. But surfaces are not artefacts of intelligence. It would add nothing to the understanding of them or to the process of designing new forms that might take to say of them that they did. Surfaces are merely surfaces.

Not all of the things we have explored are equally useful in the attainment of intelligent behaviour in the home, needless to say, nor do they achieve equal ends. The ways in which people use fridge surfaces do not reflect the fact that they orient to their daily tasks as if all were of equal importance or merit. Quite the contrary: what one sees on almost any fridge door is how some things matter more than others; how some things will matter tomorrow but not today; and how other things do not matter at all and yet are thoughtfully placed there for everyone to see. Sentiment is a form of thought too.

One could approach this diversity as a problem, one that computers could help solve. The smart home programme, as we see it, has been preoccupied with elaborate technologies to monitor human movement, the comings and goings of occupants and has sought to link this movement to various messaging systems, for example. According to this vision, the smart home will be able to check who is in the kitchen, say, and can alert that person to various messages to or to-do-items related to their being in that place.

Though this vision sounds appealing, to us it is misguided on two counts. First, and this is the weakest objection (though nonetheless a powerful one for that), we conject that this vision will be too difficult, both technologically, but also in terms of its usability. Replicating the complexity of the real world would make the system complex and vulnerable to error; it will almost certainly make it complex to use. We would expect the burden of entering in data in to the smart house to be far greater than the (unreliable) benefits that come out of it.

A stronger objection, from our view – and indeed the one we have wanted to emphasise as the leitmotif of this chapter – has to do with what one might call the balance between human and machine in this vision. This is to allude to a particular take on what intelligence might mean and imply. It seems to us, and we are strongly convinced of this, that the way people deploy their thoughtfulness at home is through steering a course between two opposites: being mechanical on the one hand and being relaxed, unplanned and almost chaotic, on the other.

To be more specific, it seems to us that what one learns from almost any ethnographic study of home settings is how things become sterile when they are routine. To send a good morning message to one's partner as they walk in to the kitchen each morning will soon become meaningless and irritating if done every day, mechanically. Tenderness between people in the home is suffocated by routine. Yet the opportunity for tenderness is squeezed by the practical requirements of living at home. There are always tasks to do and things to plan for. Though one might want to make everyday different, in practice there is a daily grind of 'housework'. This would suggest, then, that the solution is found in mixing the routine and the novel, the effortful and the relaxed in different ways. Accordingly, every person and every household is different in precisely the ways that each chooses a particular course between these opposing goals. Their choice makes each home unique. Our view is that we should design technology that allows people to make the decisions as they see fit and to reflect what they value on any particular day. Thus it is up to the individual members of a household to send a note to say they are thinking of someone else; it is up to that person to sort the digital clutter in their bowls. This is how they show their intelligence. They may do it badly, they may do it well, they may do it artfully or crudely; but this is what intelligence means. This is how we use the term; this is its correct meaning. No intelligence in the artefact for us.

This is not to say that we exclude the possibility that artefacts, in their design, cannot help people be intelligent; that they cannot, as it were, offload some of their work needed for intelligence. Certainly, we want to make some of the work easier for people but neither in the sense of 'letting the device do some of the thinking' nor in the sense of reducing the burden of choice but by making choices clearer to judge and easier to see both for those who make the choices and those who benefit from them. To be able to see at-a-glance that some one is still at work means that an individual can choose either to delay dinner or give the person still at work a call and urge them to hurry home, before it is too late. The technology here is not offering intelligence, it is only offering people in homes resources to act and think. It is this thinking, in the hearts and the minds of the occupants, that should make a home smart, not the technology used to build a home. This is our vision of the future. In this chapter, we have wanted to offer this vision as a way of looking back to the past: in to what it might have been to be intelligent in homes that have long since vanished into dust and dirt, only to be conjured with the intelligent use of the archeological imagination.

Acknowledgments we are indebted to all those who participated in our field research and prototype trials. their time, commitment and thoughts during the studies have been invaluable. Special thanks must also go to those who have worked on the design and development of the prototypes as well as helping explore and evaluate their utility. Abi Sellen, Shahram Izadi, Tim Regan and Steve Hodges played vital roles in this. Ken Wood and Lyndsay Williams also made substantial contributions during the ideation stage of the fridge magnet project. Rachel Eardley deserves particular thanks for her contributions to the interaction design for homenote, the whereabouts clock and the fridge magnet concepts.

References

Abowd G, Edwards K, Grinter B (2003) Smart homes or homes that smart? SIGCHI Bulletin. Supplement. Interactions 2003:13.

Aldrich F K (2003) Smart homes: past, present and future. In Harper R (ed.), Inside the Smart Home, Springer, London, 17–39.

Bonanni L, Lee C, Selker T (2005, April 2–7) Attention-based design of augmented reality surfaces. CHI '05 Extended Abstracts on Human Factors in Computing Systems, Portland, OR, 1128– 1231.

Button G, Coulter J, Lee J, Sharrock W (2005) Computers, Minds and Conduct, Polity Press, Cambridge.

Crabtree A, Hemmings T, Rodden T (2003) The social construction of displays. In O'Hara K et al. (eds.), Public and Situated Displays: Social and Interactional Aspects of Shared Display Technologies, Kluwer, The Netherlands, 170–190.

Edwards K, Grinter R (2001) At home with ubiquitous computing: seven challenges. In Abowd G D et al. (eds.), Proceedings of the 3rd International Conference on Ubiquitous Computing (Atlanta, GA, Sept 30–Oct 02, 2001), Springer-Verlag, London, 256–272.

Harper R (2003, September 8–11) People versus information: the evolution of mobile technology, keynote address. Mobile HCI '03, Udine, Italy, 1–18, Springer, Dordrecht.

Harper R, Shatwell B (2003) Paper-mail in the Home of the 21st Century: An analysis of the future of paper-mail and implications for the design of electronic alternatives. Journal of Interactive Marketing 3:311–323.

Hutchins E (1995) Cognition in the Wild, MIT Press, Boston.

Ju W, Hurwitz R, Judd T, Lee B (2001, Mar 31–April 5) CounterActive: an interactive cookbook for the kitchen counter. CHI '01 Extended Abstracts on Human Factors in Computing Systems, Seattle, WA, 269–270.

Kim S-H, Chung A, Ok J-H, Myung I-S, Kang H J, Woo J-K et al. (2004) Communication enhancer—appliances for better communication in a family. Personal and Ubiquitous Computing 8:221–226.

Laerhoven K V, Villar N, Schmidt A, Gellersen H-W, Håkansson M, Holmquist L E (2003) Pin&Play: the surface as network medium. IEEE Communications Magazine 41:90–96.

Norman D A (1988) The Psychology of Everyday Things. Book The Psychology of Everyday Things, Basic Books, New York.

O'Hara K, Harper R, Unger A, Wilkes J, Sharpe B, Jansen M (2005, April 02–07) TxtBoard: from text-to-person to text-to-home. Conference on Human Factors and Computing systems, CHI 2005, Portland, OR, 1705– 1708.

Petersen M G, Krogh P G, Ludvigsen M, Lykke-Olesen A (2005, April 02–07) Floor interaction HCI reaching new ground. CHI '05 Extended Abstracts on Human Factors in Computing Systems, Portland, OR, 1717–1720.

Sarvas R, Oulasvirta A, Jacucci G (2005) Building social discourse around mobile photos: a systemic perspective. 7th International Conference on Human Computer Interaction with Mobile Devices & Services, Salzburg, Austria, 31–38.

Sellen A, Harper R, Eardley R, Izadi S, Regan T, Taylor A S et al. (2006) Situated messaging in the home. Conference on Human Factors and Computing systems, CSCW 2006, Banff, Canada.

Tolmie P, Pycock J, Diggins T, MacLean A, Karsenty A (2002, April 20–25) Unremarkable computing. Conference on Human Factors in Computing Systems, CHI 2002, Minneapolis, MN, 399–406.

Chapter 7
In Context: Meaning, Materiality and Agency in the Process of Archaeological Recording

Thomas Yarrow

Text and Context

This chapter is concerned with archaeological 'context sheets', pro forma documents that are central to the way in which most British archaeological sites are recorded.[1] Whilst ostensibly unremarkable, they pose a number of interpretive challenges. As with documents more generally, the representational logic they engender seems to direct attention away from their own material and artefactual properties. Conceived as passive representations of the material properties of 'the site', their own material properties are paradoxically obscured, along with their capacity to cohere and elicit a variety of actions. Moreover, as literal embodiments of 'context', they seem to confound the anthropological assumption that context somehow exists outside or beyond the text itself.

In a discussion of the concept of the 'archaeological record', Edgeworth (2003) highlights the conceptual difficulty entailed in simultaneously imagining texts as both artefact and meaning:

> A written record can be perceived either as a material object in its own right, or as the vehicle for subjective meanings, depending on the point of view of the perceiver. It is very difficult, however, to hold both points of view simultaneously. Focus attention on the objective shape of the printed letters and their meaning disappears. Shift attention

[1] The interests that animate this chapter developed during a period of six months spent working for a commercially funded archaeological unit, where I was part of a team of approximately 12 archaeologists who excavated a large Roman and Neolithic site in Norfolk. Through this, I became interested in the role that recording sheets played in coordinating and articulating a diverse range of people and things on site. Whilst my main concern is with the use and understanding of context sheets in the practices of the archaeologists I worked with, my account also draws on subsequent analysis of context sheets taken from other sites excavated by the same unit. For reasons of confidentiality, the site and all quotes remain anonymous.

T. Yarrow
Department of Social Anthropology, School of Social Sciences, University of Manchester, Manchester, UK
e-mail: thomas.yarrow@manchester.ac.uk

C. Knappett, L. Malafouris (eds.), *Material Agency*,
DOI: 10.1007/978-0-387-74711-8_7, © Springer Science+Business Media, LLC 2008

> back to meaning of the words and their objective form becomes 'fuzzy', then recedes out of direct consciousness. (2003: 4)

This paradox, I suggest, stems from the assumption, foundational to western thought that artefacts are an inevitable and natural substrate that has to be related through systems of knowledge and categorisation. In this vein, as Strathern (1990) notes, anthropologists have largely imagined material culture as the concrete counterpoint to the abstractions of culture. Thus conceived, meaning is to be elucidated by recourse to social and cultural contexts in which artefacts are embedded. These contexts in turn make artefacts themselves appear as mere illustrations of more general patterns of thought and behaviour. With this in mind, I suggest that the anthropological (and more generally social scientific) propensity to overlook the capacity of artefacts to cohere and elicit actions, stems from this tendency to reduce the 'material world' to its symbolic or meaningful content.

Documents have provided a focus of one sort or another for a number of anthropological enquiries (e.g., Cruickshank 1998, Harper 1998, Sarris 1993). In a variety of ways, these have tended to draw attention to the different meanings that may be derived from what is manifestly the same text as it enters different contexts. In this way, attention has often focused on how texts are given different meanings in different contexts. Such an approach, which has tended in a post-structuralist vein to stress the endless possibilities of meaning and interpretation, has often overlooked the way in which the material and patterned properties of texts frame and contain such interpretations. Thus the work of the sociologist Smith (1990) introduces a necessary corrective, elucidating how texts may themselves facilitate the construction of particular versions of reality. With reference to scientific texts she describes how:

> The materiality of the text (the printed paper, the TV screen, the computer monitor) is key to the socially organized transition from paramount reality to the domain of scientific theorizing or other textually grounded domains. (1990: 81)

Although this approach conceives texts in more active terms, it shares a common assumption that texts are self-evidently set apart from the social reality in which they participate. Whilst texts may impinge upon 'paramount reality', they are not themselves a part of it.

By contrast, I borrow from the work of the anthropologist Riles who argues for the need to examine documents as a subject of enquiry 'in their own right' (Riles 2006a, b). By this, she means to draw attention to the aesthetic, artefactual and patterned qualities of documents as opposed to the capacities they have to convey 'meaning' in a more restricted sense. Crook (2007) makes a complimentary point in his elucidation of the 'textual person'. By demonstrating the person-like relations of texts and the text-like relations of persons, he reveals the limitations of analyses that take for granted a distinction between a writing subject and a textual object. The suggestion here is that texts, themselves composed of persons, animate relationships in much the same way as more conventional kinds of 'person'. In different ways, both demonstrate the importance of taking seriously the forms and relations that texts engender and the capacity for these to animate social and material relations in their own right. From such a perspective, the material and aesthetic properties of texts are not separated from the meanings they may acquire.

Critical Perspectives

In conventional understandings of fieldwork, context sheets are used to record things after the act of interpretation is complete (e.g., Roskams 2001). In other words, although they record the material form of the site and archaeologists' interpretations thereof, they do not themselves determine or affect these. By contrast, Lucas (2001a) suggests that context sheets determine the way in which archaeologists excavate and consequently the very form that the site takes.

In this analysis, attention is drawn to the normative aspects of these documents. In a Foucauldian vein, this line of enquiry suggests that particular kinds of categorisation lead to the formation of particular kinds of subjects as well as to particular kinds of objects. Thus, as Lucas has it, context sheets 'control not only the record, but also the bodies who produce it' (2001: 9). If such documents act to make the objects of archaeology comparable, then they do this by making the actions of the people that use them comparable. Here then, the emphasis is on the strategic functions that these documents perform.

Chadwick (1998) and Hodder (1999) also note how such forms constrain the actions and hence the creativity of those excavating sites. Although originally and potentially empowering in the sense that they wrested interpretation and description away from the site director, Chadwick suggests that context sheets have increasingly become 'tools of a new archaeological orthodoxy' (1997: 5), facilitating the imposition of categories and procedures on the archaeological workforce. In a related way, Hodder (1999) argues that whilst such forms potentially facilitate 'multi-vocality', in practice they have often restricted interpretation to the specific parameters engendered within the form. Whilst context sheets thus constrain those who use them, they are also represented as curtailing the capacity for the archaeological record to contradict and hence change the thoughts and ideas of those excavating. For both Chadwick and Hodder (cf. Adams and Brooke 1995), critique of conventional context sheets leads to the call for new methods of recording. Contexts sheets, they stress, should be less prescriptive and give greater emphasis to interpretation.

Important as such critiques have been for the development of new methodologies, my own concern is not with the strategic use of such forms but with the more ethnographic question of how these are used and understood by a variety of people in the practice of excavation. In eschewing a more critical perspective, my intent is to use context sheets and the documentary practices through which they are authored to reflect upon a set of debates concerning the relationship between 'materiality' and 'agency'. As my account will demonstrate, both of these terms are analytically problematic precisely because they are ethnographically significant.

Materiality and Agency

Miller (1987, 1998) was one of the first anthropologists to insist on the importance of focusing on the material world as an antidote to the pervading sociological determinism within anthropological thinking. His approach has been foundational to the development of the now well-established field of material culture studies. In a variety of ways, this has acted to counter the anthropocentric concerns of mainstream anthropology. Miller, for example, has argued against a Durkheimian tradition of thought in which social distinctions are simply projected onto the world, suggesting instead that sociological categories are made and substantiated through material culture. In this view, social and material worlds are constituted through a Hegelian dialectic: people construct themselves and their social relations through the consumption of various forms of material culture, whilst material culture is in turn transformed in line with the ideas and identities of those who consume and appropriate it.

Yet as Ingold (2000) amongst others notes, the very idea of material culture rests on an ontological separation between the material or natural world on the one hand, and the cultural world on the other. Whilst each may shape the other, the assumption of an irreducible difference entails the idea that each is characterised by fundamentally different laws. Ingold suggests that the mistake has been to confuse the many separations between different kinds of *materials*, with a fundamental separation between materiality and the mind. Thus rather than seeing action as arising out of a dialectic between the material and the social, he suggests the need to focus on the diverse kinds of matter that intersect and shape one another in the course of social life.

Whilst Ingold (2007) explicitly distances himself from aspects of the approach, the philosopher of science, Latour (1987, 1993, 1999), similarly calls into question a distinction between 'the social' and 'the natural' as an analytic starting point. In line with actor-network theorists more generally (e.g., Grint and Woolgar 1997, Hassard and Law 1999, Law 1994), he suggests that rather than pit material and social agents against one another, it is necessary to look at the hybrid networks of people and things in which different kinds of 'actants' are conjoined. Action is thus seen to emerge not in the dialectic of material and social agencies but in the ways in which different kinds of people and things are co-mingled. For example, he argues in the case of the 'sleeping policeman'[2] that various people's wills, acts and ideas are conjoined with those of gravel, concrete and paint. As such, some of the characteristics of policeman become pavement, just as some of the characteristics of pavement become policeman.

[2] Concrete or tarmac ramps built across roads with the intention of slowing traffic.

The point, as Latour notes, is not to extend subjectivity or intentionality to the material world and nor is it to treat humans as if they were objects. Rather the rejection of a subject/object leads to the position in which agency (action, intention, will) no longer has to be located by reference to one or the other of these poles. Indeed the notion of 'actants' as opposed to 'agents' is intended to signify a shift away from the assumption that intention or will are the exclusive properties of people.

The work of actor-network theorists is useful to the extent that it makes visible the contingent associations of people and things though which various kinds of ideas and actions emerge. Yet the result is both a unifying theory of things and a unifying theory of action. As others have objected (Henare et al. 2007; Strathern 1996, 49), the consequence is to flatten the diverse ways in which other groups of people might themselves imagine what a 'person' or a 'thing' is and to submerge the issues of how agency is indigenously conceived. As these recent critiques have pointed out, issues of agency and materiality cannot be resolved in advance of the ethnographic encounter. In this vein, my own account seeks to interrogate the different ways in which archaeologists themselves perceive their own and others' agency in relation to the documents through which excavation proceeds.

Archaeological Context

In order to locate my discussion of the way in which archaeological recording sheets are used and understood on site, it is necessary to understand the archaeological concept of 'context'. Archaeological 'context sheets' are integral to the way in which most modern archaeological sites are recorded in the UK. In a recent archaeological fieldwork manual, Roskams (2001) states that the objective of archaeological recording is:

> To split the site into its component parts – its stratigraphic units, however defined – and then remove them in the reverse order to which they were deposited, recording their physical, spatial and stratigraphic properties in the process and collecting finds from them to agreed sampling policies as one proceeds. (2001: 110)

Excavation thus conceived, is a process of taking the site apart, whilst producing a record that preserves not only the artefacts or finds, but also the various kinds of relationships or 'contexts' through which they are related. As Roskams elucidates, 'context' is an integral part of the archaeological record and central to the very definition of the discipline of archaeology:

> In contrast to random digging for treasure, the work of the field archaeologist must take place in controlled conditions, allowing for full recording of the physical character and spatial disposition of the stratigraphic units of the site. Such conditions do not merely allow better descriptions of the site features and yield large numbers of finds; they also let the excavator understand the latter's context of deposition and position in a sequence of development of the site. (2001: 153)

In the practice of excavation, the term 'context' refers to physically bounded units which are classified in terms of the actions imagined to give rise to them (Lucas 2001a). For example, a 'cut' refers to a context created by the removal of material (as in the case of the excavation of a pit), whilst a 'fill' refers to a context created through the deposition of material (e.g., the material that accumulates in a pit). As Lucas notes, this idea of 'context' is supported by a very specific kind of temporality: 'contexts' are conceived as homogenous moments, which may be chronologically connected to others, but which are themselves essentially atemporal (2001: 158)

The properties of such 'contexts' can be recorded in site notebooks and indeed this remains the normal practice in most university-funded excavation. However on commercially funded sites, this information is almost exclusively collected on pre-printed sheets, known as 'context sheets.' Changes in recording methods both reflect and enable changes in the practice of field archaeology more generally. With the professionalisation of archaeological fieldwork during the 1960 s and 1970 s, the process of recording has been increasingly de-centralised as excavators record and interpret the features they excavate. Whilst the physical singularity of the site notebook has its counterpart in a system where recording is largely undertaken by a single person (the site-director), the context sheet's literal capacity for division (each is mobile and separable) enables a system in which interpretations are arrived at simultaneously by diverse people on site.

The separation and connection of contexts at the heart of archaeological recording is similarly enabled by the material properties of paper itself: context sheets are detachable from one another and hence can be easily transported to the relevant place for recording. Yet their thinness and flatness mean that they are easily re-combinable – usually in ring binders – where they can be unified in a logical sequence.

Roskams (2001) hints at another way in which the physical properties of context sheets are more suited to modern excavation. By contrast to the smaller sites that characterised much academic excavation up to the sixties, developer-funded excavations are often undertaken on a much larger scale. As site size has tended to increase, this has favoured context sheets which, by contrast to the bounded site notebook, are potentially 'infinitely expandable' (Roskams 2003).

Filling In

The context sheet is divided into a number of distinct sections through which the description and interpretation of each context is fragmented (Fig. 7.1). The first section provides the space for information locating the context within the site and more specifically the grid/trench. This is followed by a space for 'description' of the physical characteristics of the context, in terms of criteria such as the texture, stoniness, colour and particle size. For example, a typical description of a ditch excavated on the site I worked reads, 'dark gray/black

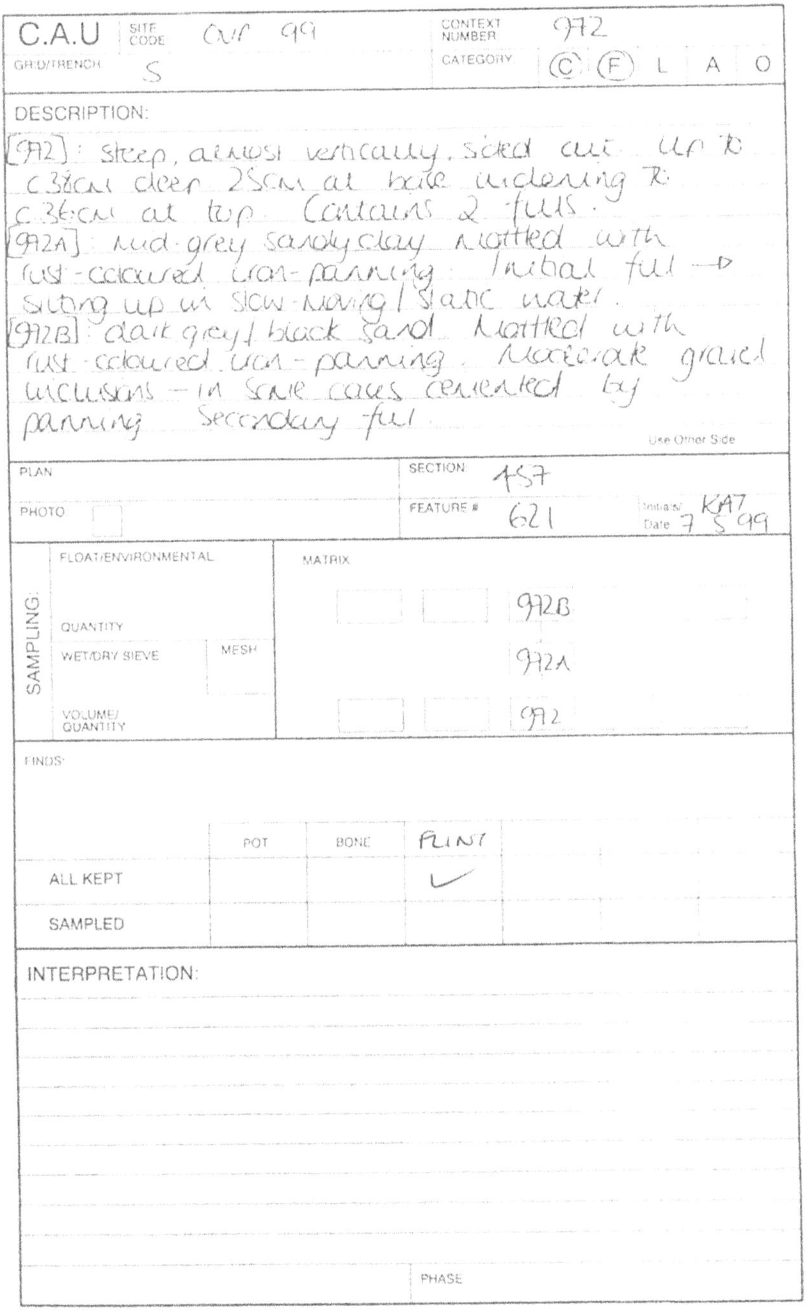

C.A.U | SITE CODE: CUR 99 | CONTEXT NUMBER: 972
GRID/TRENCH: S | CATEGORY: (C) (F) L A O

DESCRIPTION:
[972]: steep, almost vertically, sided cut up to c38cm deep 25cm at base widening to c36cm at top. Contains 2 fills.
[972A]: mid-grey sandy clay mottled with rust-coloured iron-panning. Initial fill → silting up in slow-moving / static water.
[972B]: dark grey / black sand. Mottled with rust-coloured iron-panning. Moderate gravel inclusions – in some cases cemented by panning. Secondary fill.

Use Other Side

PLAN | SECTION: 457
PHOTO | FEATURE #: 621 | Initials: KA7 | Date: 7 5 99

SAMPLING: FLOAT/ENVIRONMENTAL; QUANTITY; WET/DRY SIEVE; MESH; VOLUME/QUANTITY

MATRIX: 972B; 972A; 972

FINDS:

	POT	BONE	FLINT
ALL KEPT			✓
SAMPLED			

INTERPRETATION:

PHASE

Fig. 7.1 A typical example of a context sheet, based on a recording system developed by the Museum of London Archaeological Service (MOLAS)

sand and mottled with rust colored iron panning. Moderate gravel inclusions – in some cases cemented by panning'

Following the description of each context, the form provides a space for the documentation of each context in relation to others. Below this, a number of boxes enable those excavating to record the different kinds of artefact located within each context. Finally, a space is provided for the 'interpretation' of each context. By contrast to the 'description', which should ideally be phrased in neutral and scientific terms, this enables those excavating to make wider inferences about the likely significance of the feature in terms of the wider processes occurring on site.

The recording of context was part of a set routine. Context sheets were located in the site hut, a metal porta cabin, situated on the edge of the site, along with other recording equipment such as spare pencils and rubbers, hand tapes, 'base-plans' and permatrace for drawing plans and sections of excavated features. Tools used for excavating features such as mattocks, spades, hand-shovels and wheelbarrows, by contrast, were kept in a separate store and hence the conceptual distinction between 'excavating' and 'recording' was concretised in the physical layout of the site.

Before the recording commenced, it was a normal practice to fully excavate the archaeological 'features' (usually pits, ditches and post-holes). This was a methodical process in which attempts were made to define different 'contexts' through a variety of means. In some cases, this process of identification was relatively unproblematic and in such instances excavators spoke of the way in which the differential texture, colour or 'feel' of the soil made it easy to differentiate the extent of distinct contexts (Fig. 7.2). Yet in

Fig. 7.2 Section of a Neolithic pit. The feature contains a number of 'contexts' visible as different colours and textures in the 'fill'

other cases, differences were difficult to spot and at such times excavators employed a variety of means in order to make these distinctions apparent. For example, other people on site – specifically the site supervisor or those imagined to have more experience – were enlisted, existing 'sections' were cleaned back in order to make differences of colour and texture more apparent and in some cases people simply left the feature for a while, claiming that differences often appeared over time through the differential ways in which soil eroded.

Once a feature had been excavated and 'contexts' defined, the archaeologists would generally return to the site hut, where they would collect context sheets and sign out individual 'context numbers' from a book known as the 'site indices'. These unique context numbers meant that whilst the description of different contexts could be literally split apart and documented on separate sheets of paper, it was still possible to trace the relations between these.

The context sheet makes it possible to displace the original feature onto a form which has a number of physical properties that make it much more suitable for the task of embodying the archaeological record. Lucas (2001a) suggests that a key attribute of the archive is its capacity to be iteratively re-visited. As such, the material properties of paper are important to the role that such sheets perform in the creation of a 'record'. Unlike spoken words, paper inscriptions are able to remove descriptions from the situations in which they arise. Because of this, context sheets are able to capture particular interpretive moments, which can then be returned to by other people, at other times.

Although the distinction between 'excavation' and 'recording' was made evident in the spatial and material organisation of the site, once excavation was completed, this distinction was effectively collapsed. As Edgeworth (2003) argues, the archaeological concept of 'the record' conflates the material remains and the documents that are made of this. Thus he suggests that the text persuades both the reader and writer of its own 'reality' through a variety of rhetorical strategies. Without denying the importance of such rhetorical conventions, I would add that the physical form of the context sheet was also an important aspect of its persuasiveness as an 'objective' document of the site. Bateman (2006) describes how site plans bear the material essences of the site itself, scarred with process of their production and often stained with dirt. Similarly, context sheets, once archived, appear to retain a very concrete connection with the site, often being tattered and ripped, covered in mud and smudged by rain. In this sense, the perception that they contain information that objectively conveys the material properties of the site is enhanced by the connection made latently but very concretely in the material traces of excavation that such documents retain. Thus as Pellegram (1998) has argued in relation to bureaucratic documentary practices, the physical nature of texts has much to do with the messages they convey.

Forming Thoughts

Whilst recent archaeological accounts of context sheets have tended to stress their instrumental functions on site, a close examination suggests that they are not only or indeed primarily strategic in nature. Flicking through a ream of filed context sheets, the diversity of ways in which people fill them in quickly becomes apparent. For example, whilst some give long and detailed descriptions, others are much more cursory. Moreover, many of the forms leave large sections blank, whilst on others sections are expanded and notes or diagrams added on the back (Fig. 7.3). Sometimes the reason for leaving spaces was simply that the criteria were inappropriate to a specific feature that had been excavated, but in other cases more fundamentally important aspects of the context sheet such as descriptions, interpretations, measurements or the excavator's name were omitted. One of the archaeologists on the site on which I worked thus claimed that 'At the end of the day, they are just a piece of paper – you can fill them in how you want.' The importance of the information they carried was also confused by remarks made by a number of site supervisors, who suggested that once the site was finished these were of limited value. Most of the information on sheets was redundant, giving too much of the wrong sort of detail. In-depth description of individual contexts, it was suggested, was of little interest when it came to the interpretation of sites (Yarrow 2006a).

If context sheets are not only, or even primarily, imagined by those who use them as instruments of control, how might we interpret the importance that was nonetheless attached to them? An alternative reading is suggested by the work of the anthropologist Reed (2006), who draws attention to the way in which constraint can itself be experienced as a kind of freedom. Reed's analysis focuses on the use of the 'warrant cover', a form filled out by prison warders, in a high security prison in Bomana, Papua New Guinea, describing such things as the medical status, age and ethnicity of inmates. Whilst such documents might seem ready-made for a Foucaldian analysis, Reed suggests that a concern with their use as mechanisms of control has tended to conceal an indigenous appreciation of how these are assigned agency. From the perspective of those who use them, thoughts are imagined to be anticipated by the document's form. Since every response seems to call up the same basic design, responses simply add concrete detail to the abstract pattern already there. Prison warders thus feel themselves to be passive actors, unable to go beyond the design of the form itself. The perception, here is that agency lies not with them, but with the document's technology. Thus Reed describes how action is regarded as integral to the document's design, a movement that Bomana inmates explicitly recognise and feel led by.

Despite profound differences in terms of the epistemological schemes that support them, I suggest that context sheets were similarly regarded by many archaeologists as eliciting thoughts and responses which were not straightforwardly their own. Since every response had to conform to a strict set of textual

Fig. 7.3 Additional drawings and descriptions frequently appear on the back of context sheets. Here an elaborate diagram depicts the positioning of finds within a Neolithic pit

practices, the act of completing such forms seemed to emanate more from the document than from the person as an autonomously arrived at interpretation or description.

Perhaps this in part helps to explain the reluctance that site assistants often expressed when it came to filling them in. Where it seemed obvious what was being excavated, or where the interpretation seemed self-evident, filling in the contexts sheet was regarded as a 'pointless' or 'boring' chore. Since the descriptions always seemed 'the same', an air of inevitability seemed to surround the responses that they made. Although sometimes filled in during excavation, this was often done after the event, hurriedly and with relatively little conscious thought.

Yet if this self-generating capacity often resulted in the feeling of inevitability and in the perception that form filling simply amounted to replication, then at other times, site assistants exploited this capacity as a tool to extend their thoughts and actions. This was perhaps the case in situations where it was difficult to tell what was being excavated and where people consequently became unsure of their interpretations. In such instances, issues of 'meaning' or 'interpretation' were foreclosed through the attention that was given to 'description'. Thus, rather than considering how a particular feature was formed, site assistants would focus on the particular shape the feature took, or the kinds of fill that it contained: was it 'ovoid' or 'sub-ovoid'; was it 'flat bottomed'; was a particular fill–a-'light brown sandy silt' or a 'mid-brown silty sand'?

Sometimes those excavating would say that the context sheets 'helped them to think' about what they were digging as, for example, when it made them compare the particular feature they had excavated with others on site. In this way, the context sheet sometimes elicited particular kinds of conversations and relationships on site. Attention in these moments was often directed to the pattern or form that language took and to the comparative possibilities that it elicited, rather than to its 'meaning' as such. Form filling, I was told, had a rhythm to it as with other aspects of excavation.[3] The form seemed to pull the answers out of itself so that the response was felt to be part of the document's design as much or more than a part of their own actions. In this sense, context sheets were often seen to make visible the agency of those excavating. As one site assistant once remarked to me, 'Sometimes the fact that there's a piece of paper – that the piece of paper is actually there – reminds me that I've done something.'

Forming Identity

Context sheets were often imagined to make evident aspects of the identities of the people who filled them in. The site supervisor, who collated and interpreted the context sheets on the site on which I worked suggested that:

---

[3] Similarly Riles notes how in the context of NGO documentary practice, 'Language had a shape, a rhythm, a feel, not simply a meaning' (Riles 1998: 386).

> You can always tell from the sort of person they are how they will fill [context sheets] in. Context sheets also tell you about the people who fill them in.

In this way, he explained, the aspects of people's character that are normally hidden sometimes become apparent by looking at the context sheets they produced. By the same token, he described how people who are careful and meticulous produce careful and meticulous context sheets.

This perception that context sheets concretise the actions and thoughts of those who create them was more widely shared. Thus, it was claimed by some that unnecessarily detailed and precise descriptions were characteristic of those forms created by 'youngsters', a fact that was related variously to the way in which archaeology has become 'increasingly professionalized' and the idea that 'they've been brought up in a more objective world.'

One site supervisor suggested that variation in the way in which such forms were completed was partly related to differences in terms of knowledge of the archaeological process. Those who know how context sheets are interpretatively used, she suggested, may leave sections blank not through ignorance but because they know that certain kinds of information will not be helpful. Yet unnecessary lengthy descriptions and interpretations were not always attributed to lack of knowledge. Thus she suggested that some people write at length because they become bored with the job and feel that they are unable to express their opinions. Such people, she argued, feel empowered by the thought that their ideas are made concrete and tangible as part of a record that will endure.

While the length and content of the descriptions people provided were thus taken to evince aspects of their identity, the definition and identification of discrete contexts was also seen as a measure of personal identity. In this vein, a distinction was sometimes asserted between 'splitters' and 'lumpers': whilst the former referred to those who tended to separate contexts on the basis of small difference (leading to the identification of numerous contexts), the latter applied to those more willing to aggregate differences under a small number of contexts. The difference between 'splitters' and 'lumpers' was taken to demonstrate not only the differences of approach, but also the different personal attributes more generally. Whilst 'splitters' were imagined by some as careful and meticulous, others saw this attention to detail as 'unnecessary' or 'anal' and spoke of the way in which superfluous description was employed as a way of avoiding making interpretations or avoiding responsibility. Conversely, 'lumpers' were said by some to employ a crude and un-thinking approach, whereas for others the strategy represented a willingness to interpret what was being excavated and hence the ability to 'see the bigger picture.'

Capturing the Moment

Context sheets (as elucidated above) are widely imagined to form part of an enduring 'record'. This vision is the counterpart to the conception of excavation as destruction (Bateman 2006, cf. Lucas 2001b): the obliteration of the relations and artefacts that constitute the initial site of excavation is justified by the understanding that the record of this will endure indefinitely.

Whilst context sheets were thus formally conceived as a record of the spatial and material relations inherent in the site itself, those excavating sometimes used these sheets to capture more intimate or personal thoughts. For example, a site supervisor recounted how on one context sheet she came across the inscription, 'Dom's my best friend'. The motivation and intention behind such acts of inscription are difficult to interpret and indeed it is partly this opacity that declaration of this nature itself exploits: is it ironic or literal; flippant or serious? If interpreted as a joke, as it was by the supervisor who later came across it, then the humour seems in part to inhere in the way that an apparently highly personal expression is captured in such a manifestly objective document. The inscription seems also to flout the overt conception of context sheets as records of 'the site', by turning them into records of the people and relationships through which they are created.

Yet the supervisor who told me of this comment suggested that, while humorous, such inscriptions were nonetheless profound. In making these fleeting and incidental details concrete as part of a record that endures, people are able to, in a sense, step back and see their own actions as if through the eyes of future generations. While she suggested that the reality of excavation is often experienced as being tedious and monotonous, such inscriptions enable people to imagine what they do as belonging to something bigger than themselves.

On the back of one context sheet I came across, the site assistant who filled out the form had written:

> Remember when you find yourself in the interminable expanse of time that stretches out like vast deserts, the glorious oasis of life, the cup of tea. There will always be a brew for you with us.

It is unclear for whom this reflection is intended and perhaps it is simply a quirk of the material properties of pen and ink that has led an incidental detail to endure (much in the way that archaeological sites themselves preserve actions and traces, not conceived by those responsible for them as memorialisation). Yet the almost deliberate wording and multiple crossings out in the original text suggest a degree of care and thought that hints at an alternative reading. The author seems to have intended that their reflection should be read. If this is the case, it could be suggested that the very idea manifested in this statement was elicited by the context sheet itself and the putative audience it concretises for the site assistant. The context sheet does not simply capture a thought that already existed: it makes that thought conceivable.

Meaning, Materiality and Agency

Anthropological and social scientific theorising on the subject of materiality has often led to relatively general propositions about the nature of 'things'. In such a view, theories may illuminate the various meanings that different people might attach to artefacts in any given context, but the essential nature of those things is regarded as independent of both their theoretical and ethnographic designations. Anthropology, in other words, has often been guilty of reproducing a specifically western vision of the world in which meaning and materiality are self-evidently independent.

By contrast, if we take seriously the recent suggestion that 'things *are* meanings' (Henare et al. 2007), then a theory of things – or, indeed of 'materiality' – becomes a pointless goal. If things and meanings are inextricable from one another, then it is impossible to assert in universal terms what a 'thing' is. The issue of how artefacts are defined as against 'people' and, indeed, as against one another, will always be a matter of ethnographic interpretation.

To take the context sheet as an example, how and in what sense could this be considered 'an artefact' independent from the various people and things that contribute to its creation? Is 'it' only the paper and ink that constitutes any given sheet or the more general Platonic form that is reproduced for use on different sites? In what sense can we assert a separation between the sheets that are filled out on site and the circuitry of the computer and printer through which these are created? And how, if at all, can we disentangle such recording mechanisms from the actions and thoughts through which recording takes place? In one sense, we cannot and the conclusion to be drawn is that the assumption that human subjective agents are self-evidently distinct from the material world must always be precisely that: an assumption. Analytically speaking, it is impossible to divorce 'the context sheet' from the wider networks of human and non-human 'actants' (Latour 1999) in which it is enmeshed. In this sense, both action and meaning do not reside in the human 'subject' but are distributed in relation to a variety of people and things.

Yet archaeological recording itself attempts to bring precisely such an opposition into being (see also Yarrow 2003, 2006b). The creation of 'the record' requires that context sheets be considered as artefacts independent from the people who produced them. As such, archaeological practice attempts to resolve the issue of what is 'material' and what is only 'mental'; what is 'objective' and what 'subjective'. In doing so, it reproduces a specifically western conception of the individual, corporeally bounded by the extent of the body. The process of excavation is imagined by the archaeologists who engage in it to generate thoughts and feelings as well as stories, anecdotes and knowledge, yet these are not regarded as altering the fundamental biological or physiological attributes of the person. Hence in creating objects, archaeological recording creates the subjective agent as the largely non-discursive side effect.

The idea that things and meanings should not be separated at an analytic level does not therefore equate to an assertion that this opposition does not exist ethnographically. Rather I suggest that it is only possible to appreciate the ways in which people and things are combined or separated, if such oppositions are not assumed at the outset. Whilst archaeologists frequently assert and demonstrate the objectivity of the artefacts and contexts they unearth as distinct from their own subjective interpretations, the work required to achieve this distinction is not reducible to the opposition itself.

Acknowledgments In writing about archaeological fieldwork, my 'informants' have also been some of my most astute and perceptive critics. In particular, I wish to thank Chantal Conneller, Anwen Cooper, Duncan Garrow, Mark Knight and Lesley McFadyen, who have all contributed to the text at a number of levels. They are not, however, responsible for any shortcomings in the argument itself. The article was written with the support of a Leverhulme Early Career Research Fellowship.

References

Adams, M., and Brooke, C., 1995, Unmanaging the past: truth, data, and the human being, *Norwegian Archaeological Review* 28:93–104.

Bateman, J., 2006, Pictures, ideas and things: the production and currency of archaeological images. In *Ethnographies of Archaeological Practice: Cultural Encounters, Material Transformations*, edited by M. Edgeworth. pp. 95–102. Altamira Press, Lanham, MD.

Chadwick, A., 1998, Archaeology at the Edge of Chaos. Further towards reflexive excavation methodologies, *Assemblage Online* (http://www.shef.ac.uk/assem/3/) 3.

Crook, T., 2007, *Exchanging Skin: Anthropological Knowledge, Secrecy and Bolivip, Papua New Guinea*. Oxford University Press, London.

Cruickshank, J., 1998, *The Social Life of Stories*. University of Nebraska Press, London.

Edgeworth, M., 2003, *Acts of Discovery: An Ethnography of Archaeological Practice*. Archeopress, Oxford.

Grint, K., and Woolgar, S., 1997, *The Machine at Work: Technology, Work and Organization*. Polity Press, Cambridge.

Harper, R., 1998, *Inside the IMF: An Ethnography of Documents, Technology and Organizational Action*. Academic Press, London.

Hassard, J., and Law, J., editors, 1999, *Actor Network Theory and After*. Blackwell, Oxford.

Henare, A., Holbraad, M., and Wastell, S., 2007, Introduction: thinking through things. In *Thinking Through Things: Theorising Artefacts Ethnographically* edited by A. Henare, M. Holbraad and S. Wastell, pp. 1–31. Routledge, London.

Hodder, I., 1999, *The Archaeological Process: An Introduction*. Blackwell, Oxford.

Ingold, T., 2000, *The Perception of the Environment: Essays in Livelihood, Dwelling and Skill*. Routledge, London.

Ingold, T., 2007, Materials against materiality, *Archaeological Dialogues* 14:1–16.

Latour, B., 1987, Science in Action. *How to Follow Scientists and Engineers Through Society*. Harvard University Press, Cambridge, MA.

Latour, B., 1993, *We Have Never Been Modern*. Harvester Wheatsheaf, London.

Latour, B., 1999, *Pandora's Hope: Essays on the Reality of Science Studies*. Harvard University Press, London.

Law, J., 1994, *Organizing Modernity*. Blackwell, Oxford.

Lucas, G., 2001a, *Critical Approaches to Fieldwork: Contemporary and Historical Archaeological Practice*. Routledge, London and New York.

Lucas, G., 2001b, Destruction and the rhetoric of excavation, *Norwegian Archaeological Review* 34:35–46.

Miller, D., 1987, *Material Culture and Mass Consumption*. Blackwell, Oxford.

Miller, D., 1998, Why some things matter. In *Material Cultures: Why Some Things Matter*, edited by D. Miller, pp. 3–21. University of Chicago Press, London.

Pellegram, A., 1998, The message in paper. In *Material Cultures: Why Some Things Matter*, edited by D. Miller, pp. 103–20. University of Chicago Press, London.

Reed, A., 2006, Documents unfolding. In *Documents: Artefacts of Modern Knowledge*, edited by A. Riles, pp. 158–80. University of Michigan Press, Ann Arbor.
Riles, A., 1998, Infinity within the brackets, *American Ethnologist* 25:378–98.
Riles, A., 2006a, [Deadlines]: Removing the brackets on politics in bureaucratic and anthropological analysis. In *Documents: Artefacts of Modern Knowledge*, edited by A. Riles, pp. 71–94. University of Michigan Press, Ann Arbor.
Riles, A., 2006b, Introduction: in response. In *Documents: Artefacts of Modern Knowledge*, edited by A. Riles, pp. 1–40. University of Michigan Press, Ann Arbor.
Roskams, S., 2001, *Excavation. Cambridge Manuals in Archaeology*. Cambridge University Press, Cambridge.
Sarris, G., 1993, *Keeping Slug Women Alive: A Holistic Approach to American Indian Texts*. University of Arizona Press, London.
Smith, D., 1990, *Texts, Facts and Femininity: Exploring the Relations of Ruling*. Routledge, London.
Strathern, M., 1990, Artefacts of history: events and the interpretation of images. In *Culture and History in the Pacific*, edited by J. Siikala, pp. 25–41. Finnish Anthropological Society, Helsinki.
Strathern, M., 1996, Cutting the network, *JRAI (n.s.)* 2:517–35.
Yarrow, T., 2003, Artefactual persons: The relational capacities of persons and things in the practice of excavation, *Norwegian Archaeological Review* 36:65–73.
Yarrow, T., 2006a, Perspective matters: traversing scale through archaeological practice. In *Confronting Scale in Archaeology: Issues of Theory and Practice*, edited by G. Lock and B. Molyneaux, pp. 77–87. Springer, New York.
Yarrow, T., 2006b, Sites of knowledge: different ways of knowing an archaeological excavation. In *Ethnographies of Archaeological Practice: Cultural Encounters, Material Transformations*, edited by M. Edgeworth, pp. 20–33. Altamira Press, Lanham, MD.

Chapter 8
The Neglected Networks of Material Agency: Artefacts, Pictures and Texts

Carl Knappett

Synopsis

I argue, following Actor-Network-Theory (ANT), that agency is a process distributed across collectives of humans and nonhumans. These collectives can be considered in terms of networks, composed of heterogeneous nodes and links. Yet despite its name, Actor-*Network*-Theory has paid relatively little attention to the spatial and organisational structures of these human-nonhuman networks and their effects upon network 'behaviour' or dynamics. I draw upon some new network concepts in an attempt to fill this gap, and demonstrate my approach using an archaeological case study, one that explores the differential role of artefact, picture and text in actor networks. One reason for choosing such a case study is that archaeological approaches to agency remain anthropocentric, despite the material basis of the discipline, and have not as yet made much systematic use of ANT. Not only can archaeology benefit from ANT in tackling agency (particularly when supplemented with network concepts), but it can contribute to wider debates on agency thanks to its material basis.

Trimarans and Guns

In 2004–5 Ellen Macarthur stunned the world by sailing round-the-world at the age of only 28. This was a solo voyage completed in a mere 71 days. Phrased in this way, the virtuoso yachtswoman is clearly the 'agent' in control of this endeavour, with the yacht as her tool, responding to her skilful choices and actions. However, the 75 foot B&Q trimaran is no straightforward tool, but an ultra hi-tech piece of seafaring equipment; so much so that it can sometimes give the impression of sailing itself, with Macarthur as little more than a privileged passenger. From this perspective, perhaps the trimaran is the agent, doing all the sailing. And it is very easy to fall into a debate over what or who the real agent is: yacht or yachtswoman.

C. Knappett
Department of Art, University of Toronto, Toronto, Ontario, Canada
e-mail: carl.knappett@utoronto.ca

C. Knappett, L. Malafouris (eds.), *Material Agency*,
DOI: 10.1007/978-0-387-74711-8_8, © Springer Science+Business Media, LLC 2008

A more familiar and polarised debate of this kind is that between the pro-gun and anti-gun lobbies. While the anti-gun lobby argues that *guns* kill people, the pro-gun lobby maintains that *people* kill people, with the gun as nothing more than a neutral tool. The dualism here is between *materialist* and *sociological* explanations – the former portrays the gun as the responsible agent, the latter puts all responsibility in the hands of human agents.

Is there any way out of these dualistic debates? Is there a solution which admits that both human and gun may to some extent act, or that both Macarthur and the trimaran played a part in the round-the-world project? One solution that seems to work at first is to say that, yes, tools of this kind can have agency of a kind, a special kind of agency that is secondary. A gun has agency because human designers have intentionally delegated it with agency in their absence. Gell (1998) has argued as much for landmines and various other kinds of artefact, in particular artworks. It is agency by association; the argument is that agency has to emanate from humans. However, we should note that in this kind of secondary agency, in which primary human agency is delegated or transposed onto materials, the primary agency or 'authorship' may be obscured; and this property may come to be deliberately exploited.

There is, though, another answer. If we go back to the yachting example, it is interesting that Macarthur invariably speaks not of 'I' but of 'we'. The impression is not one of a virtuoso yachtswoman acting with considerable help from her hi-tech tools, but something much more like an equal partnership between woman and machine. Perhaps we can understand what is going on in this case with a little help from Latour, who comments not on this directly but on the question of agency with the man-gun example (Latour 1996). His approach is what is often termed 'symmetrical' – one cannot assume primacy for either man or gun. Rather, the two are mutually constituted, each being transformed by the other in their conjunction. Trying to decide whether one or the other is the agent makes little sense from this perspective. Think not of agents as entities, but of agency as a process. More specifically, agency is a process unfolding in given situations or activity frameworks and for this one can turn to the work of Kirsh (1995) and Goodwin (1994) in distributed cognition.

However, with the cases mentioned above, the situation or activity framework is relatively straightforward: there is a clearly definable task at hand, which is to be achieved by one human in conjunction with one tool. This is, in fact, often the kind of scenario selected for analysis by those working in distributed cognition: Kirsh and Maglio's Tetris, for example (Kirsh and Maglio 1994). The single artefact can, though, be deceptive. In the case of the Macarthur trimaran project, a significant web of human and technological support lies behind it, with 'Team Ellen' ('Offshore Challenges Sailing Team') being 35 strong and covering communications technology, boat design and performance and even marketing and sponsorship aspects. The agency in this project cannot be confined to the situations in which Ellen and her trimaran find themselves while at sea; it spills out across these widely distributed networks.

Actor-Network-Theory

It is this kind of phenomenon that is the meat and drink of Actor-Network-Theory,[1] of which Latour is one of the principal protagonists (e.g., Callon 1986; Law 1992; Latour 2005). Actor-Network-Theory, or hereon ANT, was devised as an approach to social phenomena that decentred the human subject, seeking to overcome the assumed ontological primacy of humans by adopting an analytical impartiality (Ashmore et al. 1994, 735). This impartiality allowed the focus to fall on objects ('nonhumans') as well as people in social collectives, rather than prejudging what should or should not belong within what we label society (Latour 2005, 72). ANT also shifted the focus onto relations – the 'semiotic' connections between the diverse elements in socio-technical ensembles. In science and technology studies, there are now many examples of how ANT has encouraged a focus on the spiralling networks of connection that seem to spread out from what appeared to have been singular, bounded technologies (see numerous papers on ANT resource, hosted by Centre for Science Studies at University of Lancaster and managed by John Law).

If we were to consider briefly the Macarthur-trimaran scenario from an ANT perspective, then the focus would fall very much on both the conjoined human-nonhuman character of the project and the multiple connections that hold it together across many scales. These connections in the Macarthur-trimaran actor network might be considered as semiotic, or syntactical; or, as Law (2000) puts it, in relation to his work on 15th–16th century Portuguese vessels:

> Hull, spars, sails, stays, stores, rudder, crew, water, winds, all of these entities (and many others) have to be held in place, so to speak *functionally*, if we are to be able to point to an object and call it a *ship*

Latour has described such objects as 'immutable mobiles' (Latour 1990; Law 2000). In other words, the semiotic / functional network has to remain immutable in order for them to move around in physical space. Law has used the example of Portuguese vessels to think through the existence of objects in different spaces, which he calls Euclidean and network spaces, although one might also term them physical and relational spaces. He argues that "objects are always performed in a multi-topological manner, and are dependent for their constancy on intersections between different topoi" (Law 2000, 7).

Yet such explicit considerations of the topologies of actor networks are surprisingly rare in ANT (see also Murdoch 1998; Latour 2005, 128–31). Despite the important advances it has kickstarted, I would argue that ANT does not go far enough in its exploration of network structures at either the micro- or macro-levels and the potential impact of these structures on network 'behaviours'. Furthermore, the different kinds of actors or actants, especially

---

[1] Here Actor-Network-Theory is fully hyphenated, following Latour 2005, but in previous incarnations it has been partially hyphenated as Actor-Network Theory, or not at all – Actor Network Theory.

when 'nonhuman', have not been detailed; what are the roles played in actor networks by, for example, artefacts, texts and pictures? I now deal with these two categories in turn – networks and actors.

Neglected Networks

Although the term 'network' is integral to 'Actor-Network-Theory', it has been surprisingly overlooked, at both micro- and macro-levels.[2] If humans and nonhumans are assembled together in complex collectives, then presumably these must have some kind of network structure, albeit shifting and dynamic. Although by no means numerous, some scholars working with ANT have highlighted the multiple topologies of the social; Mol and Law (1994), for example, focused attention in particular on 'fluid space', a move paralleled by Lee and Brown (1994) in their call for a consideration of 'smooth space' or 'rhizomatic networks', the latter drawing upon Deleuze and Guattari (1988). This is presumably in response to a perceived bias in ANT towards what Lee and Brown (1994) call 'striated space' or 'arborial networks'. As much as anything else, these efforts may be aimed against the idea that social collectives are hierarchically structured. However, there appears to be an implicit assumption that complex structures must be either chaotic or commanded, heterarchical or hierarchical. This polarity echoes that encountered in the natural sciences between regular networks on the one hand, of the character of a crystal lattice, for example, and random networks on the other (e.g., gases).

Yet social networks are rarely either regular or random, falling instead somewhere in between these two extremes. Such networks can have complex structures and dynamics, requiring particular methodologies for their study. While sociologists have long suspected that social networks have such characteristics, it is only recently that the mathematics of such networks has caught up. The main instigators of this new flurry of activity in complex networks are Duncan Watts and Steve Strogatz, who have published a series of influential articles in *Nature* and *Science*, spawning a whole field of network science across physics, biology, economics and sociology, among other disciplines (Barabási 2002; Watts 2003; Watts 2004; Evans 2005; Newman et al. 2006; Lane et al. in press).

We should perhaps not be overly surprised that this field has not intersected with ANT. But social network analysis using more established mathematical techniques has been around in sociology for a long time (Wasserman and Faust 1994; Scott 2000; Carrington et al. 2005). It may be that ANT has not intersected with this branch of sociology because the actors in these networks are invariably human, with little room for the nonhuman. Whatever the reasons, surely an

---

[2] although see Bennett 2005 for a fascinating perspective, not situated obviously within the ANT tradition but nonetheless related. Bennett draws on Deleuze's notion of 'assemblage' to examine the North American blackout of 2003 as an example of distributed human-nonhuman agency at a macro scale.

approach that takes all the pros of ANT and integrates it with a stronger 'network' perspective, one that can tackle structure and topology more systematically, might give us a better way forward for getting to grips with the ways in which agency is exercised /distributed in 'collectives' of humans and nonhumans.

It is not only at the macro-level that ANT is relatively weak with regard to network analysis. At the micro-level too, there is much more that might be done to characterise the variable connections that bring humans and nonhumans together in collectives. Latour (2005, 72) says that things can authorise, allow, afford, encourage, permit, suggest, etc. (and note the link Latour makes with Gibsonian ecological psychology here); and this is perhaps getting towards these micro-connections. But why not try to systematise these relationships more tightly and look at the kinds of connections that might occur – such as directionality, frequency, fidelity and distance (Knappett in press)? And as ANT has claimed to have a semiotic dimension in its attention to connections, why not even look at the semiotic links between entities? However, ANT may have been hamstrung in this by its adherence to a Saussurean semiotics (Watts, this volume);[3] whereas using a Peircean 'semeiotic' offers far more potential when it comes to understanding significative relationships in material culture (Knappett 2005; Preucel 2006; Watts, this volume). If parameters such as resemblance, contiguity, factorality, causality and convention could be used to analyse network properties, then perhaps we might be able to say something new about network 'behaviour' in terms of how agency is distributed across different 'nodes', human and nonhuman. However, I do not intend to pursue these points on network structure and behaviour much further in this chapter, as I explore these elsewhere; my principal concern here is with the variable character of material 'actors'.

Material Actors: Objects and Things

When it comes to the 'actors' in ANT, we have perhaps seen more on their differentiation, not just into humans and nonhumans, but actors and actants. Yet arguably ANT has not seen a concerted effort to examine the qualities of different kinds of artefact. One straightforward yet very useful distinction that we might introduce is between 'objects' and 'things'. These two terms are usually employed interchangeably in sociology, anthropology and archaeology, with little thought to their potential differences. In an attempt to underline the relationality of material meaning (and thus very much in line with the aims of ANT), Bill Brown has, in the context of literary criticism and cultural theory, proposed such a distinction (Brown 2003; see also Mitchell 2005, Schwenger 2006). Things, he argues, are ambiguous, undefined. They have a metaphysical

[3] Within the broadly Saussurean tradition, it is the work of Greimas that seems to have had the most influence upon Callon and Latour. See Latour (2005, 54, fn. 54), and Czarniawska and Hernes (2005, 7–8).

presence. It may be difficult to define 'a thing' – it may be un-nameable – for example, 'pass me that green thing over there' – the green thing is unintelligible in some way. As for objects, they materialise out of the amorphousness of things. Objects are named, understood and transparent. Object and thing thus clearly have a relational component in their definition – what is an object to one beholder might very well be a thing to another. Objecthood and thingness are relational registers.

Interestingly, a distinction between object and thing as two kinds of register has also been drawn recently by Gosden (2004). He defines objects as items that are alienable, quantifiable and disembedded from social relations. Things, on the other hand, are inalienable, possessing unquantifiable qualities and are embedded in social relations. Things exist in assemblages, in artefact communities, from which they are difficult to extract without losing much of their meaning. Things cannot be singled out, objects can.[4] There do appear to be significant overlaps between the formulations of Gosden and of Brown, and these can probably be attributed to a shared genealogy that can be traced back to Heidegger (see Harman 2005, 2007, on Heidegger's conception of object and thing).

As an aside, we should be aware of the dangers in defining objects or things in purely relational terms. This kind of approach is termed by Brown (2001, 7) "a new materialism that takes objects for granted", and is essentially what arises out of Appadurai's *The Social Life of Things*, and more fully developed in the wide-ranging and influential work of Danny Miller and colleagues on the consumption of material culture (Appadurai 1986; Miller 1987; 2005). That is to say, the emphasis is very firmly on the subject-object relation, with little attention afforded to the materiality of the object/thing itself (Watts 2007). This is reminiscent of the Latour – Lemonnier debate over the man-with-gun (i.e., Latour looks principally at relations, Lemonnier demands that the gun-ness of the gun receives attention too). The way in which 'materiality' approaches within material culture studies have overlooked the material properties of things/objects is also the subject of a powerful critique by Ingold (Ingold 2007).

Transformations: Artefact/ Image/ Text

How do things become objects or vice versa? Well, the process may not involve image and text at all but may occur when 'smooth coping' is disrupted. Here we may bring in Heidegger, as discussed by Wheeler (2005). Normally speaking, we encounter equipment as things – a hammer, for example, is 'ready-to-hand', part of a total actor network. If it breaks, however, the actor network is disrupted and the 'immutability' challenged (as it would be if Macarthur's trimaran keel snapped). At this point, the hammer (or keel) is no longer ready-to-hand but is 'present-at-hand'

[4] This distinction between thing and object is paralleled in the distinction drawn by Mitchell (2005) between totem on the one hand and fetish on the other.

(Wheeler 2005, 128–144). This is when its material properties may become more transparent, consciously recognised and named. It has, in other words, shifted from thing to object; but if quickly fixed can return seamlessly to thingness.

But what I particularly want to look at here are the ways in which images and texts might alter the status of artefacts. Gosden talks of the process of display in museums as one means of singling out or objectifying artefacts (Gosden 2004). A similar kind of process is discussed by Mitchell, who observes the process whereby found objects (or more properly, things) are turned into artworks. He describes this as a process of making an *image* of the object/thing; one example given is Jeff Koons' work *New Hoover Deluxe Shampoo Polishers* (Mitchell 2005, Fig. 33). The ordinary thing is transfigured, yet its ordinary status is never quite forgotten; one could argue it continues to haunt the image. This could describe what happens in museum display too – an *image* is made of the thing through display and in the process becomes objectified. But the artefact's thingness never quite goes away.

If an artefact can be transformed from thing into object through imagining/ imaging, then might words have the same effect? Peter Schwenger certainly argues as much, pointing to the ways in which words can make objects of (artefactual) things. The act of naming something with a word makes it into an object and nullifies its 'thingness' – what Schwenger calls, rather dramatically, 'the murder of the thing' (Schwenger 2001; 2006). Naming a bowl a bowl or a dog a dog establishes a lordship over it and denies its thingness. Thingness is somehow beyond representation and is thus unavoidably transformed in the act of objectification. The word denies thingness in much the same way that the image does. In both cases, a process of categorisation means that we close ourselves off to things – cognition overrules the senses (cf. Brown 2003).

Schwenger is careful to point out, however, that (artefactual?) things are *transformed* in this process rather than *annihilated.* Murder may bring an end to the physical thingness but not necessarily the metaphysical. The spectre of the thing may live on:

> If there is a murder of the thing by the word, then, this does not definitively annihilate that thing; it only transposes it to the scene of an interminable haunting of language. (Schwenger 2001, 113)

The original thing, unrepresented, is always there trailing the representation like a shadow, whether that representation be an image or a word. But what is of added interest is that both image and word do very often have a material, artefactual existence themselves, in picture and text respectively. But when the word is made text, does it continue to exist solely as an object, or might it also 'lapse' back into thingness? In that, objecthood and thingness are registers which might equally 'afflict' artefact, picture and text; then surely a textual artefact can become a thing. Indeed, Bill Brown argues that the text is striving to become a certain kind of thing, rather than a representation of things.

There is, potentially, a fascinating temporality at work here – whereby an (artefactual) object can, through display, naming, or imagining /imaging, be brought out of its latent thingness, but that once there it requires the play of a set of forces to keep it suspended in objecthood and prevent it from being drawn back into the soup of thingness. Furthermore, the naming or imaging of the thing as object may itself have an artefactual dimension, with image becoming picture, much as word becomes text. And then picture and text do not remain as objectifiers, but themselves are brought into thingness. And then those doing the naming may themselves be named: a recursive, reiterative process.

This approach is very much consistent with the aims of ANT, in its twin focus on associations and on materiality. By looking in a little more detail at some of the different registers materiality can take – that is, objecthood and thingness – I hope to develop a deeper understanding of the overall character of actor networks as human-nonhuman collectives. In particular, I focus on the transformations that can occur in these networks as artefacts are imagised or verbalised, and as images and words are in turn artefactualised as pictures and texts respectively. These changes are almost incessant in social collectives of various kinds and they alter the texture of the network and hence not only the overall dynamics, but also the particular agency of individual objects and things, as these are relationally derived.

This interdisciplinary move is a difficult one, as it takes us into visual culture and literary theory; but this foray is worthwhile for what it can provide with regard to 'material agency'. And fortunately there are scholars in these areas now exploring the relationship between texts and artefacts (e.g., Brown 2003) and images and artefacts (e.g., Mitchell 2005; also Gell 1998; Renfrew 2003; Renfrew et al. 2004; Gosden 2004). What I would like to do next is introduce an archaeological case study in which the interactions between these three categories – artefact, image and text – can be traced over space and time. These three – let us say material, visual and textual culture – are rarely considered together within a single methodological framework, as indeed is illustrated in our case study.

Artefact, Picture and Text in the Aegean Bronze Age

A deep-rooted separation of material, image and word – of artefact, picture and text – is pervasive in Aegean Bronze Age archaeology. Texts (largely in the form of clay documents) tend to be treated by specialists, often with more epigraphical than archaeological training. Cretan Hieroglyphic, Linear A and Linear B, only the last of which is actually deciphered, requires philological skills well beyond most prehistoric archaeologists. 'Pictures' of various kinds, on wall paintings, seals and pottery, are also the subject of a particular iconographical/art historical approach which often fails to tie in effectively with archaeological approaches to artefacts. Despite these methodological barriers to a conjoined study of artefact, picture and text, they need to be confronted here if we are to gain some impression of their status as things and/or objects.

The Aegean Bronze Age is actually a highly suitable domain of enquiry because of the wide range of artefacts, pictures and texts in various media.

While artefacts and images are of course common throughout the Aegean Bronze Age, texts are not, with the first writing on clay documents not appearing until the early part of the 2nd millennium BC on Crete. Cretan Hieroglyphic and Linear A are the two earliest scripts known, the latter gradually supplanting the former. Linear A then disappears at the expense of Linear B, probably some time in the 15th century BC. Linear B, an early form of Greek, is the only one of these scripts to have been deciphered.

These three scripts are used in various ways in relation to artefacts and pictures. We will first consider the relationship between text and picture, before moving on to that between text and artefact.

Text and Picture

Although Aegean Bronze Age scripts do have many signs which appear pictographic, all three scripts are principally of a syllabic character: Cretan Hieroglyphic has around 90 syllabograms, Linear A has 75 syllabic signs and Linear B 87 (Olivier 1986, 378–9). Nevertheless, a certain number of signs are recognised by scholars as being logograms (or 'ideograms') rather than syllabograms, and many of these are pictorial – so the logogram for chariot in Linear B is a schematic iconic depiction of a chariot. In Linear A, for example, there are 19 signs which are ideograms (or more correctly, logograms) of pottery shapes, and all of these are pictorial (Godart and Olivier 1976–1985; see Fig. 8.1). Other pictorial Linear A ideograms exist for wine and olive oil, as they do (amongst others) in Linear B. In Cretan Hieroglyphic there are currently 33 identified logograms, and in some of these cases syllabic signs double up as logograms, a phenomenon also seen in Linear A and B (Olivier and Godart 1996, 16).

There is clearly a rich interplay at work here between artefact, picture and text, extremely valuable for our current focus on how they interrelate in the transformation of things into objects. Pictographic representation in texts, while perhaps murdering the thing through the word, does maintain a shadow of the 'murdered' thing. In the transformation of the representation from pictograph to ideograph, the image may, of course, be read less and less as an icon and more and more as a symbol, with the shadow hence gradually fading. And here I think we can very usefully go a step further by following Robertson (2004) in distinguishing between *direct* and *indirect* representation. Indirect representation is when an icon that looks like the object represented is used to reference the word. The representation is thus indirect because it proceeds from iconic sign to object to word. Direct representation, on the other hand, is when the textual signs correspond to syllables or phonemes – the link between the sign and the spoken sound is direct.

With indirect representation, the textual sign is identifying with the object/thing; it is as if the textual sign is in sympathy with the 'murdered' thing it references (the image of the thing retains its memory). This kind of sign

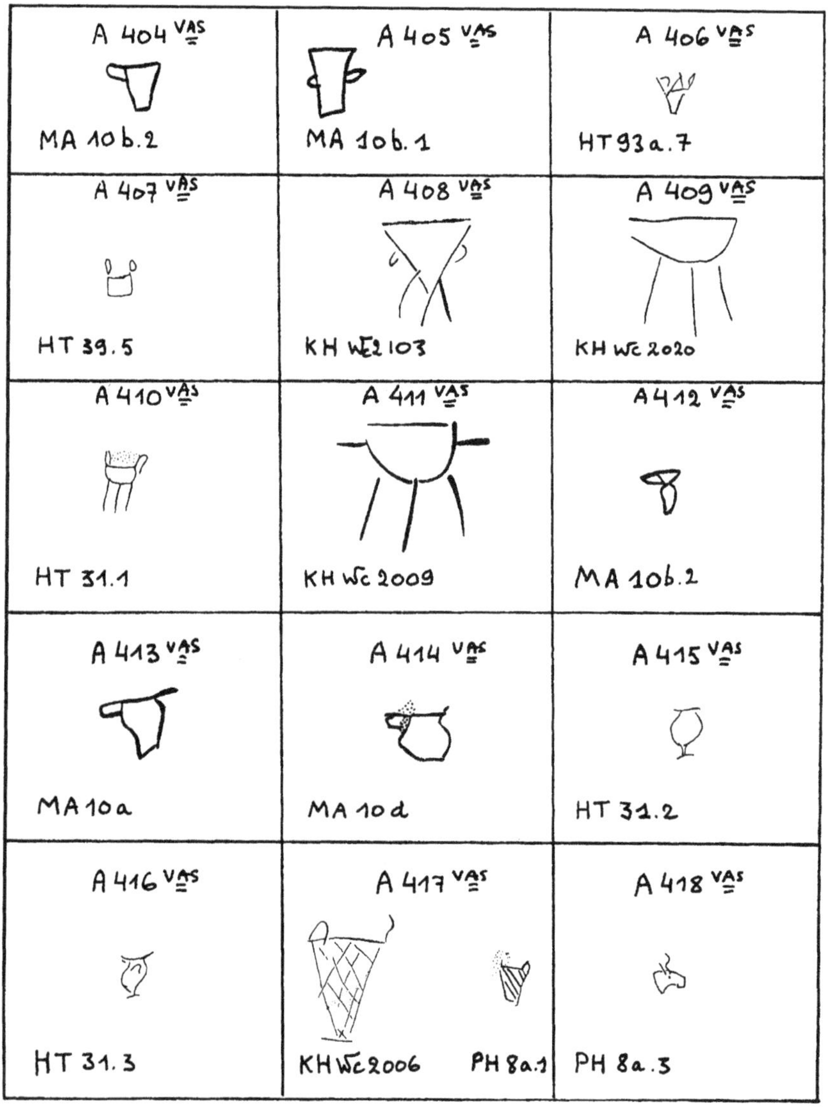

Fig. 8.1 Some Linear A logograms of pottery shapes (from Godart and Olivier 1976–1985, vol. 5, LII)

establishes a relationship of resemblance with the referent, rather than sacrificing that iconic link in favour of an auditory one. Robertson argues that *indirection* is typical of the first writing; he also observes that this does create difficulties, as many concepts do not lend themselves to simplistic iconography (e.g., how do you draw the word 'for'?)

However, a sign looking like it might be an image of something does not mean that it is necessarily *functioning* as a pictograph. Pictographs can become ideographs – which is to say that the sign ceases to signal its referent through visual resemblance, and instead operates through a habitual and conventional association. Further still, the process of representation can become rather more direct, with a visual sign becoming a syllabogram, that is, taking on a phonetic value. Robertson suggests that this might often happen through 'acrophony', that is, by taking the first syllable of the word represented by a logogram. So if olive oil is 'elaiwon' and is represented by what looks like an olive branch, then at some stage in script development the olive branch may come to stand for the first syllable of the word 'elaiwon', for example, 'el'.

So, what is the situation with respect to the Aegean scripts of Cretan Hieroglyphic, Linear A and Linear B? We have already mentioned that while many of the signs may appear to be pictorial, most are syllabic signs. Applying Robertson's terms, each script appears to function through direct representation. Here one might raise another point of interest from Robertson: that a general process of change seems to be at work in writing systems, such that a progression from icon to symbol (from indirect to direct) is seen, but rarely vice versa. Is there any indication of a progression from indirect to direct representation in the Aegean Bronze Age scripts, as Robertson would expect? The simple answer seems to be 'no', unless we are simply missing the earlier script or scripts out of which Cretan Hieroglyphic and Linear A evolved. This is not impossible, as the origins of each script are unclear. One observation we might add, however, is that there does seem to be a certain degree of indirect representation in Cretan Hieroglyphic. If we look particularly at Hieroglyphic seals, Hieroglyphic signs are used in conjunction with decorative motifs on very small sealstones (often only 1–2 cm across; see Fig. 8.2).

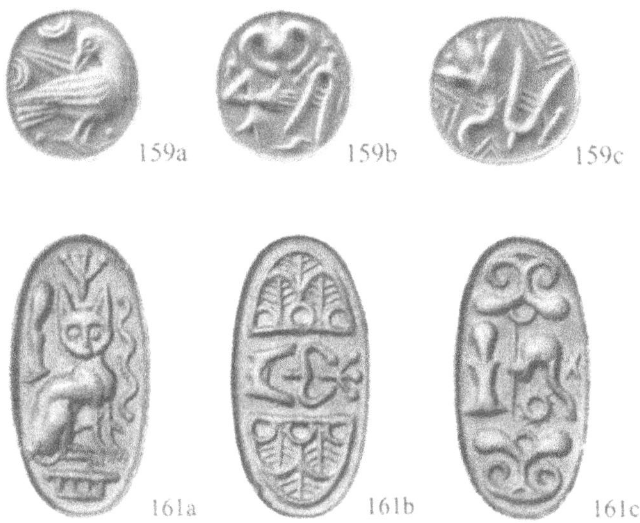

Fig. 8.2 Sealstones with Cretan Hieroglyphic (from Krzyszkowska 2005)

This raises the question as to whether these signs were really meant to be 'read' as signs, or whether their pictorial qualities were in such instances more significant (see Krzyszkowska 2005, 95–8). We might consider this some indication of an early use of indirect representation, one that fades over time. When considered in the light of Schwenger's comments above, an evolution from indirect to direct representation can almost be seen as a gradual dispelling of the 'shadow' of the object. Yet this process never quite seems complete as long as the syllabograms maintain an iconic, imagistic aspect.[5]

Text and Artefact

While above we have given some thought to the various ways in which artefacts may be represented pictographically and textually, we might also consider the relationship between text and artefact much more directly. The question we can ask is this: how closely associated are texts and the artefacts they reference? Especially in Cretan Hieroglyphic and Linear A, textual documents can be very closely associated with the commodities they name – in the form of what are called 'direct object sealings' (Weingarten 1986; Schoep 1999; Krzyszkowska 2005, 99–101). These sealings consist of lumps of clay pressed over the mouths of jars and pithoi, or over pegs securing chests or doors, and then stamped with seals (Fig. 8.3). Thinking in Schwenger's terms, naming and labeling may 'murder' the thing and commodify it, but the murdering text stays close to the body just to be sure. Of course, it risks being drawn back into the soup of thingness itself.

Another category consists not of sealings associated with artefacts, but of artefacts on which script has been directly inscribed or incised. This phenomenon is encountered in all three scripts. In Cretan Hieroglyphic there are very few examples, with the majority occurring on small juglets known as 'Chamaizi pots' (Olivier and Godart 1996, 294–311). These juglets have between two and five signs incised or painted on their bodies, usually at the maximum diameter (Fig. 8.4). As the script remains undeciphered, the meaning of these signs is unclear.

Linear A inscriptions on pottery are rather more frequent, with around 37 examples (Olivier 1986, 384). Many of these occur on storage jars ('pithoi'), and are often found incised close to the vessel mouth (Godart and Olivier 1976–85; see Fig. 8.5). There is a concentration of these inscriptions in one period in particular, Middle Minoan IIIB, a feature noted by Sir Arthur Evans. Some have been found beyond Crete too, with a recent example coming to light at the Cycladic site of Akrotiri on Thera (Karnava and Nikolakopoulou 2005). As with the Cretan Hieroglyphic examples, the meanings of these inscriptions are unknown; do they

[5] Whether or not the syllabograms were formed through a process of acrophony is probably a matter for debate, but they do seem to maintain some iconic memory.

Fig. 8.3 Reverse side of a direct object sealing from Malia (from Krzyszkowska 2005)

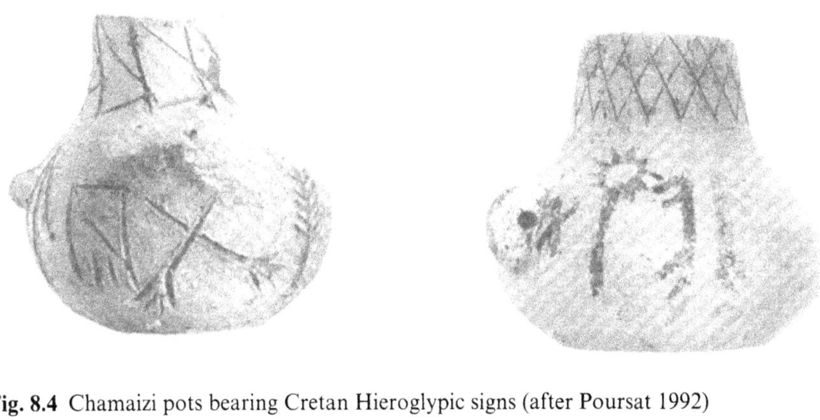

Fig. 8.4 Chamaizi pots bearing Cretan Hieroglypic signs (after Poursat 1992)

Fig. 8.5 Linear A inscription on a pithos from Akrotiri, Thera (from Karnava and Nikolakopoulou 2005)

relate to the contents of the pot, the producer, or perhaps the intended consumer? What is striking is the very different kind of vessels involved in the two scripts, with the Chamaizi vases of Cretan Hieroglyphic presumably containing very small quantities of something valuable, while the pithoi of Linear A are designed for bulk storage of commodities such as oil, wine or grain.

Turning to Linear B, problems of decoding the meaning of such inscriptions no longer apply. Furthermore, there are more examples, with Olivier recording 144 inscriptions painted on pottery vessels (Olivier 1986, 384; although by 1999, Van Alfen mentions 180 examples). As with earlier examples, the inscriptions usually only consist of a handful of signs (see Fig. 8.6), and they refer to toponyms or personal names in many cases (Van Alfen 1999). While Van Alfen argues that these inscriptions were indeed meant to be read, and played a role in the palatial bureaucracies, some scholars have argued that they may instead have had a largely decorative function (see Van Alfen 1999, 253, fn. 6). He goes on to argue that the inscribed stirrup jars themselves functioned not only as transport vessels but also as documents, playing the role that sealing nodules played in other settings. Making the artefact into a textual document is a very direct example of the role that texts can play in objectifying various categories of things.

It would appear that Aegean scripts and their documents play an objectifying role through their capacity to name and image certain kinds of thing. They do seem to be strongly focused on 'things' that lend themselves to iconic naming/imaging, primarily material commodities such as grain, oil, wool, metal etc (i.e., not many abstract concepts such as life/death/god which might be rather more problematic in their representation!). But once objectified and artefactualised do these clay documents not then become things themselves (i.e., parts of assemblages)? We should also consider the extent to which different

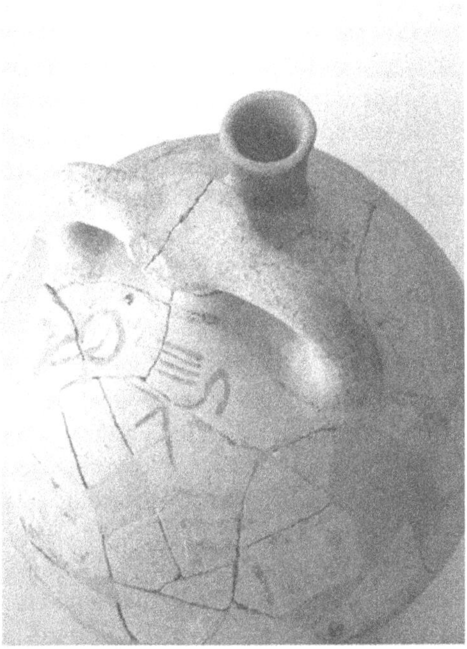

Fig. 8.6 Linear B signs on an inscribed stirrup jar from Malia

parts of a community would be able to recognise either the document itself for what it was, or indeed the signs for what they represent. Linear A documents might very well have been *objects* to some (nameable, transparent) but *things* to others (un-nameable, incomprehensible).

Discussion

We might benefit from Law's work on Portuguese vessels as a means of thinking through the interactions between texts, images and artefacts in Aegean Bronze Age contexts. We have to consider network *functionality*, particularly the overriding function of texts in administration, elites having set up a network, or 'macro actor' (Czarniawska and Hernes 2005, 9), for controlling the flow of commodities. This control is achieved, as much as anything else, by *making objects of things*. And while initially we can see the use of artefacts and images in this process, words do seem to come into the process increasingly. Moreover, images and words themselves then take changing artefactual form, with their distance from the imaged or named artefact increasing/decreasing in network and/or physical space.[6] The innumerable

[6] It might be useful here to use terms deployed by Murdoch (1998), who differentiates between spaces of prescription and spaces of negotiation. The introduction of texts into networks may serve to formalise those networks and create spaces of prescription instead of negotiation.

connections and disjunctures between artefacts, pictures and texts create particular physical and relational topologies that, to my mind, merit much further study than has been possible here in this brief treatment.

Let us now return explicitly to the theme of agency. If we consider the agency of these Aegean Bronze Age elites seeking to oversee the production and distribution of commodities, then we may observe that this agency is very much contingent upon complex networks of interconnection (see also Malafouris in press). These networks may not only have complex multiple topologies, but also may be composed of a wide variety of actants – artefacts, pictures and texts – that are assembled to create the macro actor that is Minoan administration. Therefore, a shift of emphasis is required. This need not be dehumanising or deindividualising or place undue emphasis on impersonal networks, things and objects; but it is certainly non-anthropocentric. One might have expected this kind of perspective to be welcomed in archaeology, given the discipline's inevitable material bias (at least in its methods). However, approaches from ANT or parallel domains have been only very rarely applied to archaeological scenarios, with a few recent exceptions (Olsen 2003; Witmore 2004; Webmoor and Witmore 2005; Watts 2007, & this volume). Despite its reticence, archaeology is well-placed to make a unique contribution to the widening debate on the character of nonhuman, 'material' agency.

Acknowledgments I am grateful to Chris Witmore for commenting on a draft of this chapter. My thanks also go to Olga Krzyszkowska for permission to use the images in Figs. 2–3, and to Artemis Karnava and Irene Nikolakopoulou for Fig. 5.

References

Appadurai, A., editor, 1986, *The Social Life of Things: Commodities in Cultural Perspective*. Cambridge University Press, Cambridge.

Ashmore, M., Wooffitt, R., and Harding, S., 1994, Humans and others, agents and things, *American Behavioral Scientist* 37 (6): 733–740.

Barabási, A.-L., 2002, *Linked: The New Science of Networks*. Perseus, Cambridge, MA.

Bennett, J., 2005, The agency of assemblages and the North American blackout, *Public Culture* 17(3): 445–65.

Brown, B., 2001, Thing theory, *Critical Inquiry* 28(1): 1–22.

Brown, B., 2003, *A Sense of Things: The Object Matter of American Literature*. University of Chicago Press, Chicago.

Callon, M., 1986, Some elements for a sociology of translation : domestication of the scallops and the fishermen of St-Brieuc Bay. In *Power, Action and Belief : A New Sociology of Knowledge* edited by J. Law, pp. 196–223. Sociological Review Monograph, Routledge and Kegan Paul, London.

Carrington, P.J., Scott, J., and Wasserman, S., editors, 2005, *Models and Methods in Social Network Analysis*. Cambridge University Press, Cambridge.

Czarniawska, B., and Hernes, T., 2005, Constructing macro actors according to ANT. In *Actor-Network Theory and Organizing*, edited by B. Czarniawska and T. Hernes, pp. 7–13. Liber and Copenhagen Business School Press, Copenhagen.

Deleuze, G., and Guattari, F., 1988, *A Thousand Plateaus: Capitalism and Schizophrenia*. The Athlone Press, London.

Evans, T.S., 2005, Complex networks, *Contemporary Physics* 45: 455–474.

Gell, A., 1998, *Art and Agency: An Anthropological Theory*. Clarendon Press, Oxford.

Godart, L., and Olivier, J.-P., 1976–85. *Recueil des inscriptions en Linéaire A*. Études Crétoises 21, vols. 1–5. Librairie Orientaliste Paul Geuthner, Paris.

Goodwin, C., 1994, Professional vision, *American Anthropologist* 96(3): 606–33.

Gosden, C., 2004, Making and display: our aesthetic appreciation of things and objects. In *Substance, Memory, Display: Archaeology and Art*, edited by C. Renfrew, C. Gosden and E. DeMarrais, pp. 35–45. McDonald Institute Monographs, Cambridge.

Harman, G., 2005, Heidegger on objects and things. In *Making Things Public. Atmospheres of Democracy*, edited by B. Latour and P. Weibel, pp. 268–271. MIT Press, Cambridge, MA.

Harman, G., 2007. *Heidegger Explained: From Phenomenon to Thing*. Open Court Publishing Co., US.

Ingold, T., 2007, Materials against materiality, *Archaeological Dialogues* 14(1): 1–16.

Karnava, A., and Nikolakopoulou, I., 2005, A Pithos fragment with a linear A Inscription from Akrotiri, Thera, *Studi Micenei ed Egeo-Anatolici* 47: 213–225.

Kirsh, D., 1995, The intelligent use of space, *Artificial Intelligence* 73: 31–68.

Kirsh, D., and Maglio, P., 1994, On distinguishing epistemic from pragmatic action, *Cognitive Science* 18: 513–549.

Knappett, C., 2005, *Thinking Through Material Culture: An Interdisciplinary Perspective*. University of Pennsylvania Press, Philadelphia.

Knappett, C., in press, Meaning in miniature: semiotic networks in material culture. In *Excavating the Mind: Cross-sections Through Culture, Cognition and Materiality*, edited by M. Jessen, N. Johanssen and H.J. Jensen. Aarhus University Press, Aarhus.

Krzyszkowska, O., 2005, *Aegean Seals: An Introduction*. Bulletin of the Institute of Classical Studies, Supplement 85. London.

Lane, D., Pumain, D., Van der Leeuw, S.E., and West, G., editors, in press, *Complexity Perspectives on Innovation and Social Change*. Springer Methodos series, Berlin.

Latour, B., 1990, Drawing Things Together. In *Representation in Scientific Practice*, edited by M. Lynch and S. Woolgar, pp. 19–68. MIT Press, Cambridge, MA.

Latour, B., 1996, Lettre à mon ami Pierre sur l'anthropologie symétrique, Culture Matérielle et Modernité – *Ethnologie Française* 26(1): 32–37.

Latour, B., 2005. *Reassembling the Social: An Introduction to Actor-Network-Theory*. Oxford University Press, Oxford.

Law, J., editor, 1992, *A Sociology of Monsters: Essays on Power, Technology and Domination*. Routledge Sociological Review Monograph, London.

Law, J., 2000, Objects, Spaces and Others. Published by the Centre for Science Studies, Lancaster University, at http://www.comp.lancs.ac.uk/sociology/papers/Law-Objects-Spaces-Others.pdf

Lee, N., and Brown, S., 1994, Otherness and the actor network: the undiscovered continent, *American Behavioral Scientist* 37(6): 772–790.

Malafouris, L., in press, Archaeology and distributed cognition: excavating a mind not limited by the skin. In *Excavating the Mind: Cross-Sections Through Culture, Cognition and Materiality*, edited by M. Jessen, N. Johanssen and H.J. Jensen. Aarhus University Press, Aarhus.

Miller, D., 1987, *Material Culture and Mass Consumption*. Blackwell, Oxford.

Miller, D., editor, 2005, *Materiality*. Duke University Press, Durham.

Mitchell, W. J. T., 2005, *What Do Pictures Want? The Lives and Loves of Images*. Chicago University Press, Chicago.

Mol, A., and Law, J., 1994, Regions, Networks and Fluids: Anaemia and Social Topology, *Social Studies of Science* 24: 641–71.

Murdoch, J., 1998, The spaces of actor-network theory, *Geoforum* 29(4): 357–74.

Newman, M., Barabási, A.-L., and Watts, D.J., editors, 2006, *The Structure and Dynamics of Networks*. Princeton University Press, Princeton.

Olivier, J.-P., 1986, Cretan writing in the second millennium BC, *World Archaeology* 17(3): 377–89.
Olivier, J.-P., and Godart, L., 1996, *Corpus Hieroglyphicarum Inscriptionum Cretae*. Etudes Crétoises 31, Paris.
Olsen, B., 2003, Material culture after text: re-membering things, *Norwegian Archaeological Review* 36(2): 87–104.
Preucel, R.W., 2006, *Archaeological Semiotics*. Blackwell, Oxford.
Renfrew, C., 2003, *Figuring It Out. The Parallel Visions of Artists and Archaeologists*. Thames and Hudson, London.
Renfrew, C., Gosden C., and DeMarrais, E., editors, 2004, *Substance, Memory, Display: Archaeology and Art*. McDonald Institute Monographs, Cambridge.
Robertson, J. S., 2004, The possibility and actuality of writing. In S. D. Houston, editor, *The First Writing: Script Invention as History and Process*, pp. 16–38. Cambridge University Press, Cambridge.
Schoep, I., 1999, The origins of writing and administration on crete, *Oxford Journal of Archaeology* 18(3): 265–76.
Schwenger, P., 2001, Words and the murder of the thing, *Critical Inquiry* 28(1): 99–113.
Schwenger, P., 2006, *The Tears of Things: Melancholy and Physical Objects*. University of Minnesota Press, Minneapolis.
Scott, J., 2000, *Social Network Analysis: A Handbook*. Sage, London.
Van Alfen, P. G., 1999, The Linear B Inscribed Stirrup Jars as Links in an Administrative chain, *Minos* 31–32 [1996–97]: 251–274.
Wasserman, S., and Faust, K., 1994, *Social Network Analysis: Methods and Applications*. Cambridge University Press, Cambridge.
Watts, D., 2003, *Six Degrees: The Science of a Connected Age*. Heinemann, London.
Watts. D., 2004, The "New" science of networks, *Annual Review of Sociology* 30: 243–70.
Watts, C., 2007, From purification to mediation: overcoming artifactual 'Otherness' with and in actor-network theory, *Journal of Iberian Archaeology* 9/10:39–54.
Webmoor,T., and Witmore, C., 2005, *Symmetrical Archaeology*, Metamedia, Stanford University @ http://traumwerk.stanford.edu:3455/Symmetry/home, accessed *26-11-06*.
Weingarten, J., 1986, The sealing structures of Minoan Crete: MM II Phaistos to the destruction of the palace of Knossos. Part I: The evidence until the LM IB destructions, *Oxford Journal of Archaeology* 5: 279–98.
Wheeler, M., 2005, *Reconstructing the Cognitive World: The Next Step*. MIT Press, Cambridge, MA.
Witmore, C., 2004, On multiple fields. Between the material world and media: two cases from the Peloponnesus, Greece, *Archaeological Dialogues* 11(2): 133–64.

Chapter 9
Some Stimulating Solutions

Andrew Cochrane

> *It will have blood; they say, blood will have blood: Stones have been known to move and trees to speak;*
> Shakespeare, *Macbeth*. Act 3. Scene 4.

Introduction

How do Irish passage tomb motifs influence people? What are the politics of spectators engaging with depicted motifs on stone? How might people construct cosmologies, worldviews and beliefs through these entanglements? Can broader understandings be created by including intimate engagements with essences in and of the world? Is it appropriate or indeed useful to regard *all* 'non-living' materials as inanimate and passive in contrast to the 'living' animate and active aspects of an environment? These questions form the initial stimuli for an investigation of the motifs present on some Irish Neolithic passage tombs. By further considering the dynamic effects that the application of liquids can make to images carved in to stone, I will demonstrate how modern dichotomies and certain assumptions increasingly impede richer expressions of an interpretation of a Neolithic period in Ireland. Rather than viewing the images through a textual representational analogy, I will instead utilise visual cultural, material agency and neurological studies, set within a worldview perspective to paint a picture of the possible ambiguities of life and belief at some locations in Neolithic Ireland.

To fully contextualise my position, I consider how some non-Western people do not immediately regard elements in the world as either being living:non-living, animate:inanimate or person:non-person; there is not a necessary separation between human beings and 'non-agentic' others (see Bidney 1953, 157; Hallowell 1975, 143; Latour 1993, 54; 1999, 174; Meskell 2005, 5). Things in the world can be experienced as animate or 'alive' in a variety of ways, which

A. Cochrane
School of History and Archaeology, University of Cardiff, Cardiff, Wales, UK
e-mail: cochranea@cardiff.ac.uk

C. Knappett, L. Malafouris (eds.), *Material Agency*,
DOI: 10.1007/978-0-387-74711-8_9, © Springer Science+Business Media, LLC 2008

can include social, biological, linguistic and belief systems (see Knappett 2002, 98). Here, I discuss how some people construct cosmologies and worldviews, and how these modes of thought can influence action. The images carved on the stones of Irish passage tombs (such as Knowth Site 1 and Newgrange Site 1, Co. Meath; see Fig. 9.1), and indeed the materials used to create these structures, were not necessarily regarded by some Neolithic people as static, permanent or eternal, as mere decoration without agency or a sense of volition. Rather they may have been experienced as fluid, dynamic and in some instances imbued with personhood and histories of their own. Motifs were invariably not all applied at the same time as one complete and static composition. Instead, there were episodes and sequences of the overlay and underlay of particular images that replaced or substituted existing motifs by imposed motifs (see O'Sullivan 1986; 1996; Eogan 1997; Shee Twohig 2000; Jones 2004; Cochrane 2006). Such an understanding (re)presents the decorated passage tombs stones as paradoxically resonating with permanence in flux. Visions, context, substances and belief layered together often generate ruptures in life that can facilitate new imaginings within the rhythms and sequences of images. Within such a perspective, the Irish passage tomb motifs will present fresh conditions for possibility and diverse understanding. Here, due to restrictions of space, I focus only on the influences of liquids on the final stages of motif placement.[1]

Fig. 9.1 The "three-spiral figure", orthostat C10, Newgrange Site 1, Co. Meath (photo: Ken Williams)

[1] It is entirely possible that liquids played an active role in the creation of the motifs themselves – this remains, however, a topic for another paper.

Visualising Environmental Agency

Within some archaeological and anthropological literature, there is a trend to consider the world from a non-anthropocentric viewpoint. No longer is it taken for granted that material objects are purely passive, functional and practical. Instead, some suggest that non-human objects or entities can sometimes possess agency, that is, inanimate objects can sometimes have personhood, history, dynamism and movement and play an active role in and through human relations. Gell (1998) proposed that we consider agency, intention, causation, result and transformation when engaging with imagery. By amalgamating these viewpoints and by introducing the importance of awe, enchantment, magic and partial cognitive indecipherability in imagery, Gell (1998) was able to propose common properties in a variety of examples including the *Mona Lisa*, New Zealand Maori meeting houses, Kongo 'nail fetishes' and miniature medieval cathedrals. Gell (1998) believed that he was able to move further towards providing the theoretical keys to understanding visual imagery. Material objects are stated to be the equivalent of persons or social agents, as opposed to being polarised or distinct from them. These agents operate within systems of action that 'change' the world rather than encode symbolic propositions about it (Gell 1998, 6). For Gell (1998), this is why agency is primary. Agents are defined as persons or things, which have the ability and intention to 'cause' something 'in the vicinity' or 'in the milieu' to happen (Gell 1998, 7). Gell (1998) distinguishes between primary and secondary agents, with the former being intentional beings (e.g., patrons, 'artists' or originators), and the latter being artefacts through which primary agents render their agency effective. These latter artefacts are described with the term 'index', to remove the appellation 'art' and to imply that they are indexes of agency. These indexes indicate by their existence intentions, actions and efficacy. From this perspective, visual images are 'about', and located in, the milieu of social relations, which are in turn dynamic relations of agency (Gell 1998). Utilising Gell's (1998) proposals, an Irish passage tomb design is equivalent to a person and indexes intention, agency and efficacy. The engraving need not be 'beautiful', although they generally are, nor does it only 'symbolise' or necessarily 'represent', although it can (see also Bloch 1995). Rather the passage tomb design *is*, it instantiates that to which it relates to. Broader understandings of passage tomb motifs can be created when we reconsider what they do, as opposed to what they are.

In reviewing the concept of primary and secondary agency, Gosden (2001, 164–5) argues that such distinctions perpetuate the dualism of animate people and inanimate things. He proposes that such dichotomies can be side-stepped by acknowledging that things '...are active in the manner of objects not in the manner of people...' (Gosden 2001, 165), with an ability to elicit and channelise particular sensory responses or effects, which in turn can influence social action (see also Gosden 2005, 194–7). Indeed, the concept of 'object agency' has recently been described as an oxymoronic catch-phrase (Webmoor and

Witmore forthcoming), which is as Russell suggests a 'non-statement' (2007, 73). Elsewhere, to overcome these difficulties, I have proposed that we should instead think of the elements in the world as being mixtures of mixtures. Meaning that there have never been any pure forms and that we should begin with mixtures, rather than end with them (Cochrane 2007). With respect to decorated Irish passage tombs, such an approach facilitates amalgamations of action, rather than distinguishing things as being either an object with agency or a human with agency.

In Gell's (1998) anthropological imagery nexus, four terms are employed: artist/originator, index, recipient and prototype. These four categories are described as having numerous identities. For instance, the originator can be the maker or the patron; the recipient can be the audience or the buyer; the index can be any artefact from which agency can be abducted; and the prototype can be a simulation (for an idol) or a sitter (for a portrait). Furthermore, any of these four categories can be mixed via relational ties – that is, one part acting on another. Such mixtures are not, however, stable and can invert, rupture and contest at any given moment. These dynamic relationships are deemed by Gell (1998) as essential, although he does concede that they can be unexpected. For example, just as an Irish Neolithic 'originator' may index their agency and intention in the production of a passage tomb engraving, so might the stone, through hardness and texture, dictate what would be done with it by the engraver. This is not to suggest that the motifs merely represent the agency or intentionality of a person but rather that the varied elements in the world influence and ricochet off each other. If indeed objects have agency, it is because people experience them as such through two-way processes of mediation. This observation does not perpetuate an object:person divide that is reified via dialectical bonds but rather it expresses the interweaving weft and warp of essences in the world. For such performances to perpetuate they must be contextualised within a supportive belief system; the following briefly details how cosmologies can assist in these particular experiences of being.

Cosmologies and Worldly Engagements

Cosmologies serve to orient persons in their world and inform them in the broadest terms of who they are and their relations to the rest of creation (Bowie 2000, 119). Cosmologies are statements in allegorical form about the overlapping interrelations and experiences of the world and the learned aspects of being (Ridington and Ridington 1975). People inhabit a world in which cosmologies (i.e., a totality), are explicitly and implicitly constructed, interactions created and taboos dictated often under modes of experiential surveillance. Activities that are performed within cosmological frameworks are social constructs of an objectified ideology and recently some scholars have postulated that such beliefs are the result of cognitive fluidity (e.g., Mithen 1996; 1999).

Boyer (1994; 1996), for instance, argues that this cognitive fluidity enables humans to create events and situations that defy the laws of physics such as talking trees and moving stones. Whilst they allow for objects in the world a sense of agency or personhood, Mithen's (1996; 1999) and Boyer's (1994; 1996) cognitive domain or evolutionary psychology arguments are, however, entrenched in a modernist episteme founded on scientific realism (Sahlins 1977; Rose and Rose 2000, especially chapter 15). Such cognitive ontologies separate the human mind from the body in the world (Clark 1997, 83–84; see also this volume). The body is reduced to an information delivery system for received data to be later processed by the mind, rather than engaging with the cognition itself. Instead it might be better to envisage processing loops, not residing in a domain space in the mind, within the skull, but instead fluidly penetrating both the human body and its environment (Clark 1997, 69). In emphasising this point, Clark (1997, 75) considers a scenario in which a person has to remember to buy alcohol for a party. To jog their memory, the protagonist places an empty beer can on the front doormat. Later when attempting to leave the house, they trip over it and recall their mission, thereby exploiting some aspect of the world (Clark 1997, 75). Other examples of how essences in the world can directly affect performance include road speed bumps (often termed 'sleeping policemen'), which influence the decision to reduce car velocity and keys that not only unlock the door to a property, but also force the handler to lock the door behind them in order to gain entry, such as the Berlin Key (Latour 1999, 186; 2000, 20; see also Knappett 2002, 99).[2] Another example can include beliefs in *Feng Shui* (meaning 'wind-water'), which is the Chinese practice of arrangement of objects and space to achieve harmony within an environment. In both these interactions, the varied elements in the world can determine the required social action, such as maintaining security obligations or where to position furniture and produce entangled webs in which social relations can move along. From such a perspective, the structural stones with motifs in Irish passage tombs, with differing colours and textures, may therefore also have *actively* created social interconnections within dynamic and sensuous mixtures.

Cosmology is part of an interactive system of meaningful, and sometimes meaningless, experience conducted in the context of a person's mind, body, soul and environment (which are not necessarily separate) that is both personally and socially informed (Bloch 1995; Hollan 2000). Cosmology and its associated images become animated and validated through the neuro-cognitive and embodied mechanisms that are responsible for the interpretations of reality experienced by each agent. Cosmologies may be imagined or expressed through a society's visual media which permit persons to intimately engage in their own version or interpretation of a reality. It has been suggested that people rarely

[2] Clark terms this trick of using the environment to prompt actions 'indirect emergence' (1997, 75).

create new cosmological beliefs and practices, but rather accept and participate in the trans-generational worldviews that they inherit from their society (Ortiz 1952, 135; see discussions in Russell 2004). Such a proposition does not, however, account for or explain social innovation or variation, be it intentional or through repeated acts never being identical (Rappaport 1999). Similarly, James (1995, 70) asserted that all experience is a process, no point of view ever being the *last* one.

Cosmological ideas are often emotionally reinforced and sanctioned through myth and specialised acts (although these formats can have other roles), and members who ignore the demands of the worldviews often suffer negative emotions or actions from particular agents (Overing and Passes 2000). The prime values of Ojibwa ontology are an excellent anthropological example of this (Hallowell 1955). Within this North American Indian culture, great stress is placed upon sharing what one has with others. Hoarding or any manifestation of greed is considered to violate a basic mythological and moral code and is subject to a punitive sanction. Hallowell (1975, 173) recalled how his friend and informant Chief Berens once fell ill and could not explain it. The Chief did, however, later recall overlooking one man when he had passed around an alcoholic beverage. He believed the man was offended and had cursed him. Hallowell (1975) describes this situation as demonstrating the extreme sensitivity of persons to the principles of sharing in Ojibwa worldviews. The Murngin of Australia also illustrates this point; the social context of their worldview justifies an enforcing social behaviour, with obligations and prohibitions being expressed in mythology and supported by seasonal cleansing acts (Warner 1958, 396). As such, a Murngin need not violate another person in any physical manner in order to be punished. By failing to adhere to the worldview rules of conduct, the well-being of the society is threatened and therefore the offending agent must be sanctioned (a contemporary example would be the USA Patriot Act). These brief examples serve to demonstrate how some elements of the world can *actively* create consequences for a person's existence through social experience and mediation.

Metamorphosis, Experience and Orientation

Some anthropological evidence suggests that not all societies make *permanent* polarised distinctions between particular states of being (see above). For instance, the grammatical structure of the North American Ojibwa language often utilises conflicting references. Thus the substantives for some, but not all, elements are regarded as animate, such as stones, trees, most materials, thunder and rain. Most Ojibwa do not appear to construct animate:inanimate dichotomies, or be consciously aware of these abstract notions (Hallowell 1975, 146). For the Ojibwa, although most can make distinctions between person:object, animate:inanimate and so on, the difference is that they mostly view all the elements as having the same ontological status (Hallowell 1975, 147). Such cosmologies and beliefs allow the Ojibwa to recognise fluid animacy and personhood *a priori* in all elements of the world under certain circumstances. The crucial test is experience – if a person can offer testimony to an event, be it when awake, in a trance or dream (they make no distinctions here either), then it is accepted by others. As such Hallowell was given accounts of particular stones that chased people around a tent, opened their 'mouths' or talked (1975, 148–9). Perplexed by this information, Hallowell asked an old Ojibwa man whether *all* stones about them were alive, apparently the old man reflected a long while then replied, 'No! but some are!' (1975, 147).

This proposal has recently been appraised by Ingold (2000, 97) who suggests that some objects are just thought of as more in a *condition of being* than others and they reach this state through an ongoing process of metamorphosis, orientation, renewal or refreshment (see also Hallowell 1975, 158). In essence, life is a task that needs to be worked at, in order to achieve a generation or state of being (Ingold 2000, 98). Therefore it is entirely possible that the engraved motifs on passage tombs, that we might regard as static or inert, were regarded by some Neolithic people as active with a sense of movement, if *experienced* as such within particular belief systems. I further suggest that these experiences would have been periodically enhanced, refreshed and stimulated by the application of liquids. How these engagements were thought about will be explored below.

Irish Passage Tombs

There are multiple ways of thinking about and seeing the world in which one lives. I believe that these various and sometimes conflicting perspectives were played out with and often central to the design and execution of passage tombs, their settings and their associated motifs. I will now briefly introduce two passage tomb sites (Knowth Site 1 and Newgrange Site 1) from the Boyne Valley, Co. Meath and their relationships to the motifs, the river and landscape (see Fig. 9.2).

The Irish passage tombs originated in the fourth millennium BC (e.g., the Mound of the Hostages, Tara, Co. Meath) and continued to be constructed until the early third millennium BC (Grogan 1991; Sheridan 2003; O'Sullivan 2005). Some passage tombs distinguish themselves from other classes of Irish megalithic tomb by incorporating the eponymous passage (e.g., Newgrange Site 1) and engraved imagery (Eogan 1986), while others do not have passages (e.g., Knockmany, Co. Tyrone). The visual imagery is non-representational and consists of geometric and other abstract motifs. In the Boyne Valley, the motifs occur on the kerbstones and the interior structural stones of the passage tombs, rendering it the richest area of megalithic motifs in western Europe (Shee Twohig 1981; Eogan 1986; O'Sullivan 1993). Such a wealth of visual imagery suggests that, contrary to Herity's arguments, the motifs were not a 'by-product' (1974, 107) or surplus extra designed to passively decorate the world. I propose that their importance was integral to the worldviews that helped create the monuments and subsequent encounters with them (Cochrane 2005, 250).

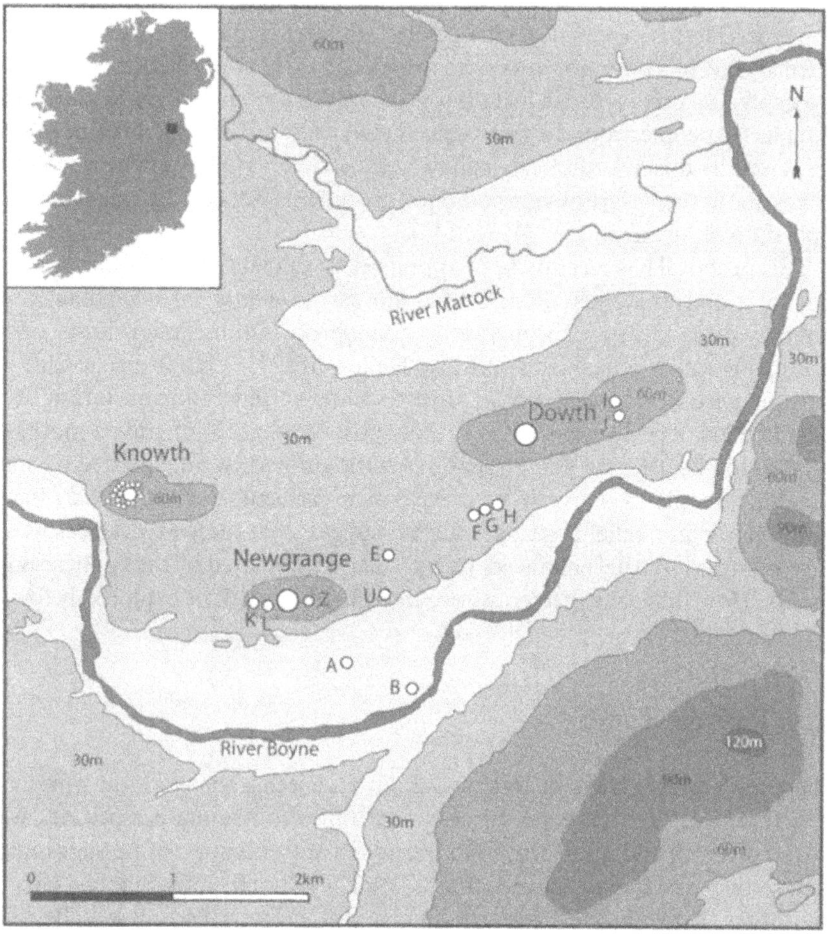

Fig. 9.2 The Boyne Valley passage tombs discussed in this chapter in their 'island' settings, Co. Meath, Ireland (adapted from Eogan 1986, 13)

The passage tombs and their motifs will not be regarded as static, but rather 'animate', fluid and prone to transformations.

Knowth Site 1 is a large mound measuring 80m by 95 m and outlined by 127 kerbstones, with two internal passage tombs (the eastern and western tombs) and it is surrounded by at least 17 smaller passage tombs (Eogan 1986). The eastern tomb is cruciform and has a passage 35 m long and the western is undifferentiated and angled with a passage 32 m long. The entrances to both passages are diametrically opposed to each other and appear to be aligned on the equinoxes (Eogan 1986, 178). It is on the kerb and interiors of Knowth Site 1 that a large quantity of motifs are found (more than 300 decorated stones have been discovered). Other visual stimuli that may have supported possible worldviews include the incorporation of white quartz and dark granodiorite in the structure, additional spreads of quartz and oval stone settings in front of the eastern and western passages, stone basins and standing stones (Eogan 1986, 47, 48). The initial construction phases are estimated as being between 3200 and 2900 BC (Eogan 1991). Newgrange Site 1 passage tomb is associated with three smaller passage tombs (M. O'Kelly 1982). The construction dates range between 3295 and 2925 cal. BC (Grogan 1991, 126). The main passage tomb consists of a kerbed (97 stones) ovoid mound (*c.* 85.3 m diameter) containing a cruciform internal tomb structure and passage measuring *c.* 19m long. Notable features of Newgrange Site 1 currently include a quartz façade and the roof-box, possibly positioned to permit the illumination of the rear chamber by the midwinter sun and communication with entities, such as the dead, 'ancestors' or local spirits (Lynch 1973, 152; M. O'Kelly 1982, 8; Sheridan 1985/6, 28).

The river Boyne is a prominent feature in the environment and is characterised by its bend as it flows eastwards below Slane, making a right-angled turn to the south. The rock against which the river is turned forms the western end of a ridge of Carboniferous shales, that lies at a height of approximately 61m above sea level, and continues east for three miles, supporting the three passage tombs, Knowth Site 1, Newgrange Site 1 and Dowth and their associated smaller tombs and features (C. O'Kelly 1971). The next turn of the river swings east, flowing parallel with the ridge and then northwards, thereby completing the Bend of the Boyne (Mitchell 1997). The Boyne River meets the Mattock river to the north, which itself originates in the Louth massif (Stout 2002), completing the feeling of an island setting. I propose that the early Neolithic people thought of or characterised the Bend of the Boyne as an 'island' (see also Powell 1938, 243; Cooney 1987, 84; 2000, 153), with cosmological significance and experience. These locations may have been thought of as where the sky, land and river conjoin; possibly creating an *axis mundi* worldview perspective at places that could be regarded as liminal (van Gennep 1960; Cochrane 2005); that is a place where certain performances can operate (Pearson and Shanks 2001, 53).

In the Boyne region, the land is punctuated by rivers, shallow areas and fords. These fords are important as they can either create aids for river crossings or barriers for river travel and as such may have been a central component of daily life (Stout 2002, 11). These rivers or 'riverscapes' may have been seen and thought of as being alive, regenerative, diverse, dangerous, contoured and imbued with cosmological significance and ambiguity. These fluid worldviews may have helped create senses of identity, temporality, place and history (see also Wilde 1849, 2–3; Cooney 2003, 323, 325). The siting of passage tombs near these waterways may also have altered the experiences of them in to places that carried connotations of journeys, transformation and death (Richards 1996), and different places, such as the sea and the mountains. Rivers can be physically felt and entered into, whilst also capable of causing injury, sickness and death (Buxton 1973, 363). Indeed, for the Saami of Finland, rivers are symbolically for the dead, as they are able to decompose, rot and putrefy life (Bradley 2000, 27); Eliade termed these types of rivers 'Waters of Death' (1964, 355). Journeys on rivers may have therefore been associated with acts of personal transformation (either alive or dead), in rites of passage (Fowler and Cummings 2003, 2). Rivers can be incorporated in to a three-tiered *axis mundi* worldview, although they are not always associated with the earth. For instance, the Mandari of southern Sudan, believe that both water and rivers are from the Sky or 'Spirit-of-the-above', as it is thought that all water was once rain (Buxton 1973, 364). Interestingly in a more modern context, a recurrent feature in English and German folklore is that the souls yet to be born, or unborn children, exist first in fountains, springs, lakes and flowing water (van Gennep 1960, 52). Indeed, rivers are often temporal, being prone to metamorphosis by shrinking and swelling at particular times of the year.

I described above the possible performances of the River Boyne as I am interested in accounting for the role that water flow may have played in altering the viewer's engagement or experience with the passage tombs. I will review how these solutions may have assisted in developing snapshots of an interpretation of reality on the surface of the stones. I follow an approach adopted by Ego (2001), who considered how some visual images on San rock art in southern Africa are influenced by liquids. In this model, the images metamorphose from a static state, to an animated form via the flow of water. The San regards rain as an entity and in its various incarnations, it has the power to transform images and precipitate experience. This chapter will investigate how this approach resonates with the evidence from Irish passage tombs and whether we can consider these data as playing an active and animated role via liquidation and sequence.

The Visual Impact of Patterns

Some previous interpretations of Irish passage tomb motifs have unfortunately lacked consideration of the visual impact of the images themselves. Traditional explanations have incorporated value judgements on the motifs' 'quality' of execution, design and aesthetic 'appeal' (e.g., Herity 1974; Shee Twohig 1981). I propose that most attempts at ascertaining complex Irish Neolithic aesthetic values from a contemporary Western perspective are, however, of limited value.[3] Instead, I consider how neurobiology can allow some universal cross-cultural generalisations to be made.[4] For instance, studies demonstrate that certain types of pattern can cause visual discomfort, optical illusions, headache

[3] The trend is to create a framework for artistic study that demonstrates relationships between the image and its social meanings (Layton 1991). This orthodox art historical application, however, informs little of indigenous and pre-Renaissance European contexts and more of Western notions of universal human 'culture specific' and 'period specific' aesthetics (Gell 1998, 3; Cochrane and Russell 2007).

[4] Such universals are viable as although the contexts, environments, political, social and economic features are undoubtedly different today then from the Neolithic, the world is still seen and created via the same neurobiological structures (see Zeki 1999; Hoffman 2000; Bailey 2005, 25; Cochrane 2006).

and dizziness. During the late twentieth century, there was a movement of what became termed 'op art', due to its optical effects. Works were produced consisting of coloured or monochrome patterns of curved or straight lines, grids and dots, which resulted in the viewer experiencing discomfort, optical distortions and systematic errors that misrepresent the outside environment to our brains. The trend was rather short-lived, yet it expedited research on the functioning of vision when exposed to extremely intense stimuli (Wade 1990; 1998; Wade and Swanston 2001).

When a spectator is subjected to 'dense' optical patterns, the neuro-visual system malfunctions. The extreme intensity of the pattern of lines overloads the contrast/orientation neurons of the primary visual cortex (termed areas V1, V2, V3, V3A, V4 and V5), causing them to 'leak' and cross-stimulate neighbouring neurons (Wilkins et al. 1984). This is termed the 'contextual effect', whereby the primary cortical neurons can be distorted when identical stimuli are detected by neighbouring neural columns. If viewed for sufficiently long periods, this effect will cause optical illusions, headache and dizziness. Furthermore, it can result in migraine and epileptic seizures in photosensitive sufferers (Wilkins et al. 1984). That images in the environment can so directly affect other elements or persons reinforces the possibility for fluid and transformative networks of relations within Irish Neolithic contexts.

Since some Irish passage tombs motifs resemble simple dense patterns, I will briefly discuss a number of observations and implications. The present preservation of the Irish motifs in their eroded form unfortunately precludes their use as a means of inducing 'op-art' optical effects, although it is possible that when freshly cut, there would have been sufficient contrast between line and space to create such effects. These effects may have been magnified if the engraving were used in conjunction with high contrast pigments. Based on red/brown lozenges being reported on a corbel over C10 in Cairn T at Loughcrew, Co. Meath, it has been argued that the motifs at the passage tombs in Ireland were originally painted and that environmental conditions in Ireland have not permitted survival (Breuil and Macalister 1921, 4; cf. Shee Twohig 1973). Interestingly, the passage tombs at Carrowkeel, County Sligo do not contain engraved motifs and Eogan (1986, 148) has suggested that this is a result of the limestone being too hard to engrave and that they may have been painted instead.[5] It is worth noting that in the Iberian Peninsula where it is relatively arid, painting does survive alongside engraved stone; as demonstrated by the black and red motifs painted on Orthostat C1, Forno dos Mouros, La Coruña, Spain and Orthostat C5, Antelas, Viseu, Portugal (Shee-Twohig 1981; Bueno Ramirez and Balbin Behramann 1996; Sanches 2006). Recently, it has been suggested that paint was possibly used to create a spiral motif on stone 11 of the Castlerigg stone circle in Cumbria, England, and that its

[5] It should be noted that engraved motifs have recently been discovered on a limestone roof-lintel of Tomb 51 (*Listoghil*), Carrowmore, Co. Sligo (Curran-Mulligan 1994, 14–15). The presence of these designs in an area previously thought not to contain motifs does not, however, diminish the plausibility of Eogan's (1986, 148) assertions.

complete disappearance since 1995 is due to natural weather conditions (Díaz-Andreu et al. 2006, 1585). Although there is no current evidence to suggest that the motifs in Ireland were painted, traces of pigment have recently been found by taking infra-red photographs of a number of decorated surfaces in the main chamber at Maeshowe, Orkney, which were previously thought to be unpainted (Bradley et al. 2000). If this new technique is applied to the Irish evidence, pigments may be found. As such, I propose that Irish passage tomb motifs did induce visual experiences, but without sufficient pigment evidence, it is unlikely that they created 'contextual effect' optical distortions on their own. For this to happen, we need to consider alternative neuro-biological encounters.

The visual images currently on Irish passage tombs are not produced via the application of pigment or paint; although it is possible that they were originally decorated or enhanced with colour. It is noteworthy that the importance of colour to past societies has only just begun to be discussed in depth, with scholars now proposing that colour impacts upon most conceptual systems and worldviews (e.g., Lynch 1998; Jones and Bradley 1999); Taçon 1999; Jones and MacGregor 2002). As with the engraving, the possible paintwork that might have adorned the stones may not have been applied to all motifs at the same time, since there may have been episodes of superimposition and more fluid sequences. If the stones were painted, we can argue that the erosion of pigment over time by the elements could have altered experiences of the sites as unchanging and timeless stone reflections into a simulation of life that time transforms (see also Cochrane 2006). These painted images may have been periodically re-applied and slightly distorted each time by different persons or agents. This would result in specific forms having two phases, erasure and metamorphosis, and in both of them the image becomes animated. The evidence from some Iberian tombs suggests that the primary colours used were black, white and red. These colours are interesting as they fit well with Turner's (1967) study of the Ndembu of north-western Zambia and their initiation rites and worldview belief systems, in which the colours black, white and red represent cosmological rivers and bodily products. Turner notes that the Ndembu sense these three colours as '...rivers of power flowing from a common source... permeating the whole world of sensory phenomena with their specific qualities...' (1967, 68). For the Ndembu, the colours provide stimuli for their interpretations of reality, while standing for some bodily performances, such as hunger, fear, aggression, submissiveness and excretory drives (Turner 1967, 89–90). Buxton's (1973, 382) account of the Mandari of the southern Sudan also suggested that the colours black, white and red are used to influence perception and created contested states that reverse a given order. The colours black and white particularly affect this reversal process, for instance black has a primary 'death' association, but in a rain context it is regarded as life-bringing, though at the same time it retains an inversed characteristic that embodies death through being a rain-storm, a killer or destroyer (Buxton 1973, 385). Tilley's investigation of some Middle Neolithic passage tombs in Southern Scandinavia, argued that the use of red ochre, white bones, the black of clay and flint for axe blades and the various colours of the stones used in the tomb construction, presented cosmological mixings of 'blood' with the 'semen-milk' of the 'ancestors' (1996, 316, 322). If the application of paint did occur at Irish passage tombs, it may have been considered in a similar way. Yet at the moment this is just conjecture and I am therefore more concerned with how one could animate images without paint; the next most obvious option to me seems water. Water as a liquid is an interesting element that requires further consideration within a Neolithic setting. In particular we should examine the ways rain and river water, both of which are abundant in Ireland, frames certain passage tomb settings and continues to influence our experiences with monuments today.

Seeing Through Sensitive Stones

Rain water marks the rhythm of the passing seasons and the blossoming of berries and fruits, with every society in north-western Europe experiencing meteorological fluctuations. With respect to some of the attributes of water, be it from the sky or the river, we should note that associations with life and liquid can transcend time, space and society – the politics of liquids. Life depends upon liquid; this is universal. From the concept of water being essential to life and the fountain of youth, to semen, milk, blood, bile and saliva and the like, a principle in some societies is that liquid means life while loss of liquid means death. Discussions with Ian Russell led me to the work of the artist Kiki Smith (e.g., *Untitled* 1987–90), who has recently explored some of these themes, presenting a series of twelve silvered glass water bottles (52.1 cm × 29.2 cm in diameter) each named with different bodily fluids, such as blood, semen, bile, tears and diarrhoea, to express some of the physical substitutes for the intangibles of certain beliefs. Contemporary Irish examples of beliefs operating in a similar manner include the wetting of the Blarney Stone, Cork, with saliva via the mouth as a process of bodily activation (i.e., the generation of eloquence or having the gift of the 'gab'). In another example, the Mandari of southern Sudan make associations between life, fertility, women and water, and these are demonstrated by the euphemism for pregnancy, which is termed *sipi i pele* 'water in the belly', while the word for semen is *kula yunusi* 'making water which promotes conception' (Buxton 1973, 369). A modern structural approach to this might suggest that 'wet' and 'dry' as an oppositional pair mean 'life' and 'death'. Therefore, liquids are living, whereas drying is dying. This may be an element to cremation activities, which was the predominant rite at Irish passage tombs (Herity 1974; Eogan 1984; 1986).[6] Although, I stress that such a simplistic dichotomy is not a universal state of affairs; for instance, for some in modern

[6] It has been noted that as many of the tomb sites are on acid soils, there may be a taphonomic bias against the preservation of inhumed remains (Cooney 1992, 130).

Western society the loss of semen is directly equated with the generation of life rather than its loss.

Yet in removing such dichotomies, we can imagine how more intimate entanglements were woven. The burning of human remains may have facilitated collective activities that involved groups of people engaging in public performances, perhaps at night, lighting up the environment. These fires may have been thought of as alive, being both gentle and comforting, but also unruly and dangerous, especially when roaring in the wind at night (Tringham 2005, 107). The fires would have enhanced sensational performances, with the colours of the flames and embers continually changing and providing an ephemeral focus. When in contact with organic and non-organic materials, dependent on degrees of heat, fires universally produce colours that include black, white and red (Borić 2002, 26), which may have linked or transcended associations with water, fire, life, death, regeneration, cleansing and transformation. Such effects may have dramatised the telling of tales and myths. The smoke may have brought tears to the eyes, with the smell impregnating the hair and clothes. The heat itself from the fire may have warmed the skin and fuelled passions via the 'sexual excitement' that fire can bring (Tringham 2005, 107).

Burning the dead can be seen to expedite the 'drying' process, the final removal of the liquid of life (Bloch 1998, 79). Interestingly, a parallel is seen with fruits, as they age they lose liquid and wrinkle. Might old age be thought of in the same terms? Certainly for the Mandari of southern Sudan, the human body is regarded as being both moist and dry at different stages of life (Buxton 1973, 363). An interesting modern example of liquid activation is derived from the Mangang ceremony in Malaysia, in which they pour Tapai, an alcoholic beverage, onto standing menhirs, to appease and refresh the spirits within (Phelan 1994). For some of the aborigines in the Kimberley region of Australia, the *Wandjina* belief system allows people to accept that some rocks represent clouds, whose power is renewed via the applications of liquids (mostly paint) (Vinnicombe and Mowaljarlai 1995, 234). In a prehistoric context, Schefer has remarked on the Palaeolithic engravings of Foz Côa in Portugal, that water not only preserves the figures but it *feeds* them periodically – it keeps them *alive* (1999, 100). Coles (2004, 204) comments that some Bronze Age images at Bro Utmark, Sweden were placed in locations that are often wet in order to make their viewing more dramatic. The notion of bathing images and keeping them alive through dynamic flows will be further discussed below.

If the stones were painted we could see how the erosion of the painted images could mark the passing of time, with the periodic bathing process by rain creating a gradual destruction and movement of the image – the antithesis of rain perpetuating life. This might generate relationships in which there are 'biographical perspectives' for the spectator, thus imbuing the stones with volition and 'life cycle rhythms' (see Gell 1992). Such actions or properties may have created a sense of history for the motifs and stones (Ingold 2007). The physical aspects of applying the paints, of touching the stones and motifs may have created tactile understandings, which vision alone cannot provide (Benjamin 1940, 233).[7] These embodied acts would incorporate the pores and skin and may have assisted in developing dynamic experiences of the motifs (see Knappett 2006, 240). Without paint we can think of the water not destroying but rather distorting, enhancing or 'enchanting' the image and creating or simulating animated qualities, producing alternative auras, possibly as a form of process, which acts to captivate a viewer (see Benjamin 1940; Gell 1998; 1999b).[8] Such displays may have been thought of as 'magical', producing effects which do not articulate to the spectators 'normal' sense of self, embodiedness, agency and being in the world (Gell 1998, 71).

In a modern Western context, one might describe such engagements with visual images as stimulating 'uncanny' or 'unsettling' rather than 'magical' sensations. Turning to other contexts, such as in Xishuangbanna Dai Autonomous Prefecture, southwest Yunnan, China, carnivalesque[9] sensibilities are played out though 'water splashing' engagements during the Dai New Year (*Sonkran*) celebrations. The Dai belief system states that water splashing is conducted when the gods visit the land to appraise peoples' sins; splashing water beseeches gods for future abundance (Komlosy 2004, 357). The water splashing practice also subverts normal codes of practice in that the young people often go 'beyond bounds' and enjoy throwing buckets of water (sometimes muddy) on people indiscriminately (Komlosy 2004, 358). Indeed, the carnival is often employed as a tactic for acceptable 'bad manners' between differing persons (Calavia Sáez 2004). In these examples not *everyone* has to invert or participate, they can watch if they like (sometimes at some personal risk),[10] while the marginal is briefly brought to the centre where it can mingle and then give way to other ways of being. Possible associations between water, the carnivalesque and some passage tombs are interesting as they allow one to imagine the settings as fluid and active, rather than as static and passive (see also Cochrane 2005).

---

[7] I acknowledge that there are many ways of engaging with motifs or structures that do not involve the primacy of vision, such as their acoustic qualities, but restrictions of space dictate that here I focus mainly on what can be *seen* (see discussions in Ouzman 2001; Watson 2001); Boivin 2004; Jones 2004; Curtis et al. 2004; Witmore 2006).

[8] The distortion and enhancement of images may also have been achieved via fire, with the flickering light creating a play of shadows (Lynch 1970, 40). This source of light may have simulated the movement or animation of the motifs (see also Sanches 2006, 135). Such illumination would work well with emphasis augmented by liquids.

[9] The carnivalesque is often created when the themes of the carnival subvert, distort and invert habitual or established society. In carnival, all that is marginalised and excluded, such as the mad, the scandalous and the uncertain, takes centre stage and liberates in an explosion of otherness. In this environment, 'negative' bodily expressions such as hunger, thirst, defecation and copulation become a 'positive' corrosive force; life enjoys a symbolic victory over death (see Bakhtin 1968; Stam 1989; Bailey 2005; Cochrane 2005).

[10] In a sense there are no 'real' spectators within carnivalesque environments as all near are immersed within it; they live in it with 'normal' life ceasing to exist during its time-span (Bakhtin 1968, 7).

The sensory qualities of particular substances seem to have played a role in determining what materials were employed in the construction of the Boyne Valley passage tombs. For example, the colour of stone seems to have been important in choosing what went where. In the Boyne Valley complex, the soft green coloured greywacke was the preferred (although not exclusive) medium for imagery carved on structural stones (O'Sullivan 1997, 28). The placement of white quartz façades, grey granite cobbles, platforms and standing stones in the Boyne valley complexes suggest deliberate combination and juxtaposition of substances of differing form, colour, texture and composition. The material aspects of the assemblages incorporated in to the passage tombs seem just as important before and after manipulation (see discussions in Ingold 2007). The colours of particular stones may have operated in similar ways as the imagery (O'Sullivan 1997, 30). As such, the coloured stones may have performed as agents of external manipulation, forming and moulding social ideals, through a complex web of relations, especially if emphasised though liquids. For instance, on K1 at Newgrange Site 1 (see Fig. 9.3) one can notice the difference between the areas stimulated by water and the areas not. This enhancement or activation through water or indeed any liquid, perhaps milk or blood, literally adds dimension to the image and makes it stand out. The water running is giving a 'suggestion' of movement. The liquid is adding dimension and illusion to the 'performance' of the image, which is objectively congealed in the stone. When produced in colour, one can notice the ways in which the water has enriched and literally brought out of the stone the red colours in the granite, exploring the surface tensions of the 'false relief' motifs.

Fig. 9.3 Notice how the motifs are enhanced and suggest movement through the action of running water: K1, Newgrange Site 1 (photo: author)

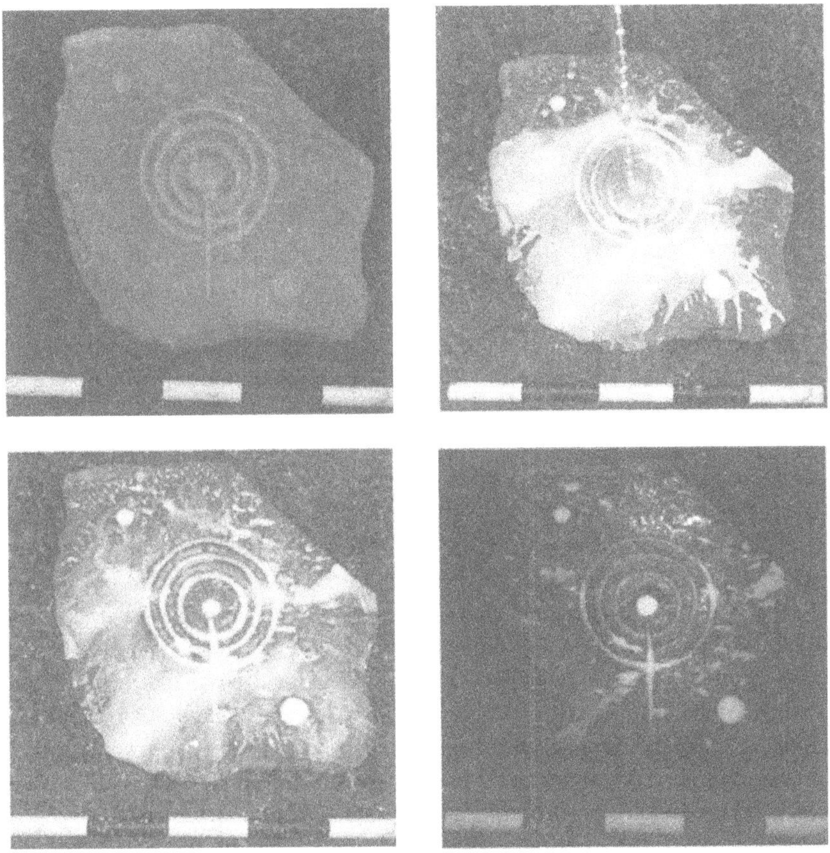

Fig. 9.4 Replica stone with cup and ring marks being stimulated by milk at Kilmartin House Museum, Argyll (photo: Author and Aaron Watson)

The water has brought to life and refreshed the engraved images and the subtleties of the stone in the viewers' minds eye, maybe vitalising the cosmological powers in the rock, defeating a static spectatorial recapitulation. The application, precipitation and evaporation of these liquids may have created a sense of temporality, (re)generation, reconfiguration and transformation (see Ingold 2007; see Fig. 9.4). Certainly, inside the passage tombs condensation caused by human breath and movement can create the impression that the stones are 'sweating' and pulsating. This affect of 'breathing' stone via human interaction in an environment can be perceptively enhanced with the usage of intoxicants or psychoactive substances. Whilst drinking coffee with Muiris O'Sullivan, he commented that when the eastern tomb at Knowth Site 1 was first opened the stones all were awash with water that had penetrated the mound, thereby exposing the varied colours of the stone – a sight that is often not witnessed today. The contrast in colour and brightness between the wet and dry states of the stone might reflect connotations

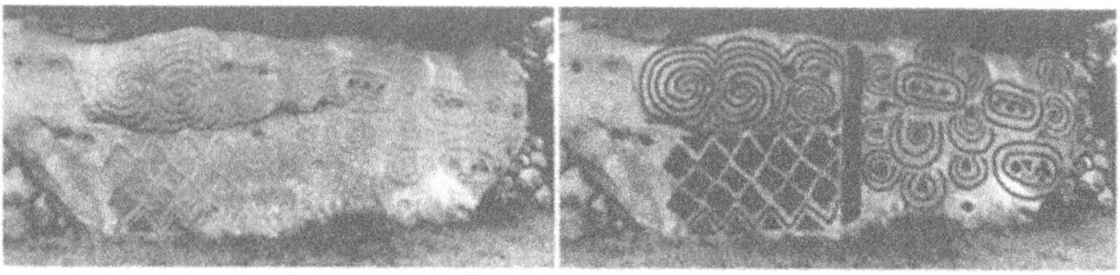

Fig. 9.5 Kerbstone 52, Newgrange Site 1 in an unpainted state and with some motifs digitally enhanced with pigment to demonstrate possible contextual effects and dense patterns – note how the stimulated images appear more vibrant (photo: author)

with the vividness of blood and the emotive and animate aspects of life (see also Cooney 2002, 101). Interestingly, K1 has a vertical channel dividing the upper half of the decorated surface in the centre. A similar channel is present on K52 that is diametrically opposite on the rear of the mound (see Fig. 9.5). This feature also occurs on the entrance stones for the eastern (K74) and western (K11) passages at Knowth 1 (Eogan 1986). At Dowth on K50 there is also a vertical line on the surface, and this stone is located directly opposite the rear of the north passage tomb, on the eastern side of the mound (O'Kelly, M. J. and O'Kelly, C. 1983). McMann (1994) argued that the vertical strokes implied that the builders believed an axial alignment extended through the structure and into the landscape, whereas Eogan (1996) has taken these vertical channels to indicate the presence of an entrance (see also Lewis Williams and Pearce 2005). This latter statement is clearly true in three examples, but is not, however, universal. I suggest the channels served a more active role, allowing liquids to flow and permeate down and through these particular and possibly significant stones.

I believe that our multiple experiences of a reality are influenced by visual images and that these constructs literally create visual or virtual realities. Whether the Irish Neolithic peoples did create complex cosmologies to experience the decorated kerbstones as being animated, momentary, fluid and flexible is something that we will never ultimately 'know'. What we can 'know' is that people, be they modern or Neolithic, depend on water and liquids for life and that people universally in temperate Europe live in situations that are periodically altered and influenced by the effects of water and rain. I am not suggesting that peoples in the past *only* experienced or used the monuments in the rain or with water, but rather that we should consider that some passage tomb settings, such as the Boyne Valley passage tombs, are framed by water and that the tombs are orientated in a context so that they repel water (i.e., with the structural corbelled roofs) and also incorporate it (such as with water flowing over the kerbstones and possibly in the basins). For instance, the passage lintels in Newgrange Site 1 had grooves or water channels cut into the upper surfaces. The grooves were up to 5 cm wide and 1 cm deep and were designed to drain rainwater outwards on to each side of the passage. Although the passage was designed to repel water, the act of entering the tomb may have been associated with plunging into the earth or even water, with the loss of light and change in temperature being immediately noticeable (Fowler and Cummings 2003, 14). It has previously been suggested that passage tombs can often amplify or distort the effects of temperature, humidity, sound and light or darkness (Bradley 1989; Watson and Keating 1999; Watson 2001). Such elemental relations are able to transcend not only the structural dichotomies of wet and dry, but also inside and outside.

Intimate engagements with these passage tombs may have stimulated emotional responses through their architecture, content and context. One may imagine the stress undergone by a person isolated in a place associated with enormous power (Dronfield 1996), such as the Newgrange Site 1 passage tombs. It has been suggested that architecture and location are a deliberate exercise in controlling and manipulating how tombs and the activities that would have taken place at them were experienced (Thomas 1990; Kirk 1993; Richards 1993; Fraser 1998). Passage tombs can therefore restrict and influence bodily movement, the senses, space, airflow and temperature. Furthermore, tombs can disorientate, humble and generate fear in those who enter, creating what Squair has called an 'architecture of inconvenience' (1998). This environment could be accentuated with repetitive sound waves produced by percussive instruments or by an infrasonic component of thunderstorms, both of which can induce altered states of consciousness (Watson and Keating 1999). It is suggested that tinnitus (a condition of ringing, buzzing, hissing or humming in the ears) from any cause, such as drumming, can trigger auditory hallucinations of music or even speech (Gordon 1993). The recent discovery of a prehistoric musical instrument constructed from six graded cylindrical hollow yew wood pipes, arranged side by side, at Charlesland, Co. Wicklow and dated between 2120 and 2085 cal. BC, does suggest an early knowledge for the generation of sound waves (in modern terms) or 'noise' (Molloy 2004). Indeed it has been proposed that no known societies, past or present, lack the ability to produce sounds at some level (Seeger 2002, 686). Altered states of consciousness can be provoked by repetitive noise in enclosed spaces, such as passage tombs or even modern small 'Drum and Bass' clubs, and these events are sometimes described as feeling like one is submerged underwater or underground (Lewis-Williams and Dowson 1990, 10–11).

Such underwater or underground description certainly fits well with the 'cavernous' or 'under-earth' architecture at Newgrange Site 1. At the deepest end of the passage is located a chamber, measuring over 5.25m by 6.5m. This is cruciform in plan, consisting of the central chamber at the end of the passage and three smaller chambers or recesses on the north (cell 2), east (cell 3) and west (cell 1). Cell 3 is the largest of the three and is the most profusely ornamented. Cell 1 is only one orthostat deep, while both cells 2 and 3 are two orthostats deep, resulting in an irregular cruciform plan. Stone basins are set on the floors of each of the recesses; there are two basins in the eastern recess (M. O'Kelly 1982, 21). The original function of these basins is unclear, but they might suggest that grinding and mixing of materials performed a role in the practices that occurred within the chambers. Some have suggested that these basins may have held materials such as cremated bone (Herity 1974, 119; Shee Twohig

1990, 43), and I suggest that liquids may have been added to the mix, possibly activating and refreshing the dry human remains. Liquids are certainly recorded at some Neolithic cruciform chambered sites; for instance at Barclodiad y Gawres, Anglesey, Wales, a liquid stew that contained wrasse, eel, whiting, frog, toad, natterjack, grass snake, mouse, shrew and hare was used to douse a fire in the central area of the tomb (Lynch 1970, 37).

The liquids may also have included milk and semen as demonstrated by the semen/milk transactions by the Sambia of Melanesia (Gell 1999a, 58–63). In these Sambia 'rituals', semen is used to activate the semen in other males, who are thought of as 'female' at this stage and the production of milk and children in females (see also Strathern 1988); liquids are therefore used actively in relations that create the generation and continuation of life and the expression of gender. Neolithic people may have regarded themselves at some level as integral wholes in which liquids could be transmitted (in life or death); yet such boundaries may still have been permeable, with substances being able to flow through and between them. These permeable or partible boundaries in flux may also have been extended to include the fragmented cremations and objects, the structure of the tomb and the motifs on it – such is the nature of mixtures. Busby (1997) has recently made similar observations in studying personhood in south India. It is noted that even though south Indian people may be sexually dimorphic, this is only regarded as a symptom of the presence of liquid substances (milk and semen) flowing and mixing within and between their bodies (Busby 1997, 270).

For the Mandari of the southern Sudan, milk is used in certain rites, first because it is white and second because it is a 'living thing' (Buxton 1973, 389). Certainly the notion of the 'semen-milk of the ancestors' (Tilley 1996, 316, 322), has already been raised in other prehistoric contexts. I suggest that it is possible that these liquids were also used literally – meaning that milk or semen was applied in practice. Furthermore, and in fitting with the archaeological data, in some past societies (such as the ancient Egyptians and Greeks) the act of 'drinking' any liquid with the dead was regarded as an act of incorporation and consequently allowed one to travel amongst the dead and out of danger (van Gennep 1960, 165). Whichever liquids were used, their inclusion as a 'magical lubricant' (van Gennep 1960, 172) might be considered to have existed in certain ceremonies, such as transitional rites of passage for persons or essences with the stone basins and associated motifs.

Fluid Structures

Quartz is a recurring visual component in the Knowth Site 1 and Newgrange Site 1 passage tombs, and may therefore have played important roles in cosmological experiences. For instance, in many societies, such as the California Yuman Indians, the Australian aborigines or Western neo-shamans and pagans, quartz is considered to be 'living' or a 'live rock', with special or healing properties, often derived from water (Vinnicombe and Mowaljarlai 1995, 240;

Fig. 9.6 'Healing Stones' – blocks of white quartz crystal for sale at the National Museum of Scotland, Edinburgh (photo: author)

Pearson 2002, 142; Wallis 2003; see Fig. 9.6). The idea that quartz objects are fluid with a suggested movement is certainly interesting. It has been noted that the stone clitter on the hilltop tors in southwest England creates visual sensation that swirls in an ever-changing pattern of trance-like forms (Tilley et al. 2000, 222). Whilst at Newgrange Site 1 I noted that a similar captivating experience could be created by staring intensely at either the reconstructed quartz façade or the quartz fragments on the ground (see Fig. 9.7).[11] Some ethnographic data suggests that some worldviews are conceived as being influenced by enchanting encounters and enveloped in luminescent light. For instance, in commenting on the importance of colour, shadow and light to some pan-Amerindians, Saunders has noted that these factors can reveal and transform the appearance of varied essences in the world (2002, 213; see also Taçon 1999). In another context, the

[11] Whether the quartz was originally presented as a near vertical façade or whether it was deposited on top of the cairn as Macalister (1939) suggested, or spread out in front as is found at Knowth's Site 1 eastern tomb (Eogan 1986, 47), has recently been questioned (e.g., Bradley 1998, 101; Darvill 2002, 82; Cooney 2006, 704; Eriksen 2006, 709). Cooney (2006) has recently suggested that the quartz was removed from the cairn slopes to create a quartz platform on the ground, where performances could have occurred.

Fig. 9.7 The quartz façade and loose quartz on the ground both creating optical effects at Newgrange Site 1 (photo: author)

last known Numic rain-shaman, Bob Rabbit from north America, used quartz crystals in his weather-control ceremonies to create water and release the spirits from within the stones (Whitley et al. 1999, 235; see also Vinnicombe and Mowaljarlai 1995, 238–40). For these people, vision is widely regarded as seeing the 'essence' rather than the surface of things; in the manner that quartz can reveal the 'sacred' glow within. The decision to incorporate 'black' and 'white' stones at Knowth Site 1 and Newgrange Site 1 might therefore present a harnessing of social and mythical significances of light, colour and matter, through their transformation in to solid forms via technical skills or technologies of enchantment (see Gell 1998; 1999b; Saunders 2002).

Bergh has proposed that the animating power of quartz lies in its ability to produce or reflect light and that this may have been seen by some as a regenerative act (1995, 153). Interestingly, the roof-box entrance slit at Newgrange Site 1 was originally closed by two blocks of quartz. The roof-box lintel RS1 produced evidence of scratch marks on its surface, suggesting that people had frequently visited the roof-box and had pulled out and pushed back the blocks a number of times (M. O'Kelly 1982, 96). This opened hollow structure may have acted as some form of interactive oracle or communicative device (Lynch 1973, 152; Sheridan 1985/6, 28). Certainly, Eliade refers to quartz as 'light' and notes that some people use it to create relations (1964, 138). The transformative power of quartz is demonstrated by the Cubeo Indians of the northwest Amazon, who insert quartz crystals into a neophyte's stomach, to activate a metamorphic process that will create a shaman from a mere layman (Pearson 2002, 142).

Interestingly, the cobbles included in the quartz façade are rock fragments, pebbles and boulders and are described as being rounded to sub-rounded in shape due to water rolling (Meighan et al. 2002, 33). Mitchell's (1992) paper on the cobbles found at the entrances to the passage tombs at Newgrange Site 1 and Knowth Site 1, concluded that most of them originated from the northern shore of Dundalk Bay. In this report, particular cobbles were given complex histories, involving water, such as pre-glacial weathering, transport by ice and glacial meltwater and exposure to marine wave action. These water worn stones were then transported 30 km to the Boyne Valley by Neolithic people. More recent examinations have discovered yet more varieties of cobbled granite and place their provenance further away in the Mourne Mountains (Wilde 1849, 191; Meighan et al. 2002, 35). Fowler and Cummings argue that the incorporation or deposition of quartz at a given stone monument may have been performed in a process of 'making it wet' (2003, 14), thus marking it out as appropriate for acts of transformation. Since the majority of Neolithic people were not buried in passage tombs (or in any other contexts that we know of as of yet), one might speculate that many deceased persons were deposited into the Boyne River. If aspects of life are seen as being transformed in to water, then it may make some sense within that context to include water in the construction of passage tombs through the use of an alternative solid material medium, such as quartz. Such beliefs may have been periodically enhanced via the application of liquids to the motifs on the structural stones of the Boyne Valley passage tombs.

Conclusions

By amalgamating visual cultural perspectives and anthropological examples with contemporary neurology, I have investigated the power that images can have. The images on passage tombs are not just passive and static forms that require the penetrating gaze of the spectator, waiting to be read with meanings comprehended. Instead, they are fluid events that manipulate and affect people who engage with them over time and space – the politics of visuality. By considering images as such, I was able to further describe the possible social relations, routines and interactions that may have occurred around them and that may have influenced how people thought. One of the difficulties in expressing how Neolithic people may have thought or acted includes the constraints of the linguistic structure that we choose to use. Dualisms seem to percolate through our understandings and expressions, while the search for meanings still dominates some Western thought (see Bloch 1995 for discussion). In order to move beyond the representational, structuralist and Platonic approaches that have been previously adopted to understand Irish passage tombs (e.g., Lynch 1973; Dronfield 1996; Lewis-Williams and Pearce 2005), I feel scholars should attempt to explore relations and experiences through alternative perspectives and more intimate engagements with the elements in the world.

The motifs on passage tombs perform by blurring and disrupting perceived boundaries between interpretations of reality and social construction. The stimulation of motifs via solutions, the enhancement of their detail, context and composition provides the *punctum,* that is, the destabilising or unpredictable effects of engagement with imagery. This means that the motifs provoke the viewer to react, thus creating unsettling, unexpected and sometimes unknown experiences (see Bailey 2005, 131). The images on passage tombs engender and connect themselves to spectators and each other through two-way intimate engagements. By considering them as such we can begin to see how the motifs perform, play and produce illusions for the viewer – they actively present rather than passively represent. These simulations not only stimulate but also create gazes or perspectives through which people can often formulate belief systems. The traits of illusion may therefore include the ability to enchant and enhance, whilst masking the disappearance of an interpretation of the world, when supplanted by another. The mixtures of images, settings and passage tombs therefore socialise people – the social is visually constructed – through their being, structure and choreography. Such performances may have created cosmologies, dialogues, myths and narratives about the uniqueness of these places. These conversations would in turn factor and texture new engagements at these sites, whilst creating fresh opportunities for experience. The acts involved in applying motifs onto stone, of possibly interacting with them in differing states of being, were therefore endeavours that disrupted, transformed and distilled the processes of experience and understanding at some Irish Neolithic sites.

Acknowledgments I would like to thank Paul Webb for managing to retrieve all the images from my corrupted 1 GB memory card. Oliver Harris, Alasdair Whittle, Doug Bailey, Muiris O'Sullivan and Kate Waddington were very kind in reading earlier drafts of this chapter. The mistakes are still mine alone. Ian Russell read this chapter and also devoted much time to discussion, and for this I thank him. I would like to thank Leontia Lenehan from the Brú na Bóinne Visitor Centre near Donore, Co. Meath for repeatedly granting access to the sites, and Steve Trick for being the designated driver who suffered hours of my continual story-telling! Ken Williams has once again generously allowed me to use images from his recent 'Shadows and Stones' exhibition (see www.shadowsandstone.com) and I thank him for his support. Many thanks are given to Sharon Webb for allowing access to the Kilmartin House Museum replica and for assisting in bathing the stone in milk! I would also like to thank Carl Knappett and Lambros Malafouris for inviting me to the TAG session, for commenting on earlier versions of this chapter, and for creating this volume.

References

Bailey, D. W. 2005. *Prehistoric figurines: representation and corporeality in the Neolithic*. London: Routledge.

Bakhtin, M. M. 1968. *Rabelais and his world*. Trans. H. Iswolsky. Cambridge: MIT Press.

Benjamin, W. 1940. The work of art in the age of mechanical reproduction. In H. Arendt (ed.), *Illuminations*, 211–44. London: Fontana Press.

Bergh, S. 1995. *Landscape of the monuments: a study of the passage tombs in the Cúil Irra region, Co. Sligo, Ireland* (Akeologiska undersökningar Skrifter no. 6). Stockholm: Riksantikvarieämbetet.

Bidney, D. 1953. *Theoretical anthropology*. New York: Columbia University Press.

Bloch, M. E. F. 1995. Questions not to ask of Malagasy carvings. In I. Hodder, M. Shanks, A. Alexandri, V. Buchli, J. Carmen, J. Last, and G. Lucus (eds.), *Interpreting archaeology: finding meaning in the past*, 212–15. London: Routledge.

Bloch, M. E. F. 1998. *How We Think They Think: Anthropological Approaches to Cognition, Memory and Literacy*. Oxford: Westview Press.

Boivin, N. 2004. Rock art and rock music: petroglyphs of the south Indian Neolithic. *Antiquity* 78, 38–53.
Borić, D. 2002. Apotropaism and the temporality of colours: colourful Mesolithic-Neolithic seasons in the Danube Gorges. In A. Jones and G. Macgregor (eds.), *Colouring the Past: The Significance of Colour in Archaeological Research*, 23–43. Oxford: Berg.
Bowie, F. 2000. *The Anthropology of Religion: An Introduction*. Oxford: Blackwell Publishers Ltd.
Boyer, P. 1994. Cognitive constraints on cultural representations: natural ontologies and religious ideas. In L. A. Hirschfeld and S. A. Gelman (eds.), *Mapping the Mind: Domain Specificity in Cognition and Culture*, 391–411. Cambridge: Cambridge University Press.
Boyer, P. 1996. What makes anthropomorphism natural: intuitive ontology and cultural representation. *Journal of the Royal Anthropological Institute* 2, 83–97.
Bradley, R. 1989. Darkness and light in the design of megalithic tombs. *Oxford Journal of Archaeology* 8, 251–9.
Bradley, R. 1998. *The significance of monuments: on the shaping of human experience in Neolithic and Bronze Europe*. London: Routledge.
Bradley, R. 2000. *The Archaeology of Natural Places*. London: Routledge.
Bradley, R., Phillips, T., Richards, C. and Webb, M. 2000. Decorating the houses of the dead: incised and pecked motifs in Orkney chambered tombs. *Cambridge Archaeological Journal* 11(1), 45–67.
Breuil, H. and Macalister, R. A. S. 1921. A study of the chronology of Bronze Age sculpture in Ireland. *Proceedings of the Royal Irish Academy*, 36C, 1–9.
Bueno Ramirez, P. and Balbin Behramann, R. de. 1996. El papel del elemento antropomorfo en el arte megalítico ibérico. *Revue Archéologique de l'Ouest*, supplément n° 8, 41–64.
Busby, C. 1997. Permeable and partible persons: a comparative analysis of gender and body in south India and Melanesia. *Journal of the Royal Anthropological Institute* 3, 261–78.
Buxton, J. 1973. *Religion and Healing in Mandari*. Oxford: Oxford University Press.
Calavia Sáez, O. 2004. In search of ritual: tradition, outer world and bad manners in the Amazon. *The Journal of the Royal Anthropological Institute* 10, 157–73.
Clark, A. 1997. *Being there: Putting Brain, Body and the World Together Again*. Cambridge: MIT Press.
Cochrane, A., 2005. A taste of the unexpected: subverting *mentalités* through the motifs and settings of Irish passage tombs, in *Elements of being: mentalities, identities and movements*, (eds.) D. Hofmann, J. Mills & A. Cochrane. BAR International Series 1437, 5–19.
Cochrane, A., 2006. The simulacra and simulations of Irish Neolithic passage tombs. In *Images, representations and Heritage: Moving beyond modern approaches to archaeology*, ed. I. Russell. New York: Springer, 247–78.
Cochrane, A. 2007. We have never been material. *Journal of Iberian Archaeology* 9/10, 138–57.
Cochrane, A. and Russell, I., 2007. Visualising archaeologies: a manifesto. *Cambridge Archaeological Journal* 17(1), 3–19.
Coles, J. 2004. Rock carvings at Bro Utmark, Bohuslän, Sweden. *Proceedings of the Prehistoric Society* 70, 173–206.
Cooney, G. 1987. *North Leinster in the earlier prehistoric period (7,000 – 1,400 bc): a settlement and environmental perspective on foragers, farmers and early metallurgists*. Unpublished PhD Thesis submitted to University College, Dublin.
Cooney, G. 1992. Body politics and grave messages: Irish Neolithic mortuary practices. In N. Sharples and A. Sheridan (eds.), *Vessels for the Ancestors: Essays on the Neolithic of Britain and Irelands in Honour of Audrey Henshall*, 128–142. Edinburgh: Edinburgh University Press.
Cooney, G. 2000. *Landscapes of Neolithic Ireland*. New York: Routledge.
Cooney, G. 2002. So many shades of rock. In A. Jones and G. Macgregor (eds.), *Colouring the Past: The Significance of Colour in Archaeological Research*, 93–107. Oxford: Berg.
Cooney, G. 2003. Introduction: seeing land from the sea. *World Archaeology* 35(3), 323–328.

Cooney, G. 2006. Newgrange – a view from the platform. *Antiquity* 80, 697–710.
Curran-Mulligan, P. 1994. New megalithic art in Sligo: yes, it is art! *Archaeology Ireland* 27(8), 14–15.
Curtis, P., Wood, J. and Feeke, S. 2004. *With Hidden Noise: Sculpture, Video and Ventriloquism*. Leeds: Henry Moore Institute.
Darvill, T. 2002. White on blonde: quartz pebbles and the use of quartz at Neolithic monuments in the Isle of Man and beyond. In A. Jones and G. Macgregor (eds.), *Colouring the Past: The Significance of Colour in Archaeological Research*, 73–91. Oxford: Berg.
Díaz-Andreu, A., Brooke, C., Rainsbury, M. and Rosser, N. 2006. The spiral that vanished: the application of non-contact recording techniques to an elusive rock art motif at Castlerigg stone circle in Cumbria. *Journal of Archaeological Science* 33, 1580–87.
Dronfield, J. 1996. Entering alternative realities: cognition, art and architecture in Irish passage tombs. *Cambridge Archaeological Journal* 6(1), 37–72.
Ego, R. 2001. A pictorial device: the dynamic action of water in a few depictions of rain animals. Trans. by Paul Bahn. *Pictogram* 12, 27–34.
Eliade, M. 1964. *Shamanism: Archaic Techniques of Ecstasy*. Princeton: Princeton University Press.
Eogan, G. 1984. *Excavations at Knowth I: Smaller Passage Tombs, Neolithic Occupation and Beaker Activity*. Dublin: Royal Irish Academy.
Eogan, G. 1986. *Knowth and the Passage-Tombs of Ireland*. London: Thames and Hudson.
Eogan, G. 1991. Prehistoric and Early Historic culture change at Brugh na Bóinne. *Proceedings of the Royal Irish Academy* 91C, 105–32.
Eogan, G. 1996. Pattern and place: a preliminary study of the decorated kerbstones at Site 1, Knowth, Co. Meath and their comparative setting. *Revue Archéologique de l'Ouest*, supplément n° 8, 97–104.
Eogan, G. 1997. Overlays and underlays: aspects of megalithic art succession at Brugh na Bóinne, Ireland. *Brigantium* 10, 217–34.
Eriksen, P. 2006. The rolling stones of Newgrange. *Antiquity* 80, 709–10.
Fraser, S. M. 1998. The public forum and the space between: the materiality of social strategy in the Irish Neolithic. *Proceedings of the Prehistoric Society* 64, 203–244.
Fowler, C. 2004. *The Archaeology of Personhood: An Anthropological Approach*. London: Routledge.
Fowler, C. and Cummings, V. 2003. Places of transformation: building monuments from water and stone in the Neolithic of the Irish Sea. *Journal of the Royal Anthropological Institute* 9, 1–20.
Gell, A. 1992. *The Anthropology of Time: Cultural Constructions of Temporal Maps and Images*. Oxford: Berg.
Gell, A. 1998. *Art and Agency: An Anthropological Theory*. Oxford: Oxford University Press.
Gell, A. 1999a. Strathernograms, or, the semiotics of mixed metaphors. In A. Gell (edited by E. Hirsch), *The Art of Anthropology: Essays and Diagrams*, 29–75. London: The Athlone Press.
Gell, A. 1999b. The technology of enchantment and the enchantment of technology. In A. Gell (edited by E. Hirsch), *The Art of Anthropology: Essays and Diagrams*, 159–186. London: The Athlone Press.
Gordon, A. 1993. Benzodiazepines and ear-tinnitus, hallucinations and schizophrenia. *Canadian Journal of Psychiatry* 38(2), 156–157.
Gosden, C. 2001. Making sense: archaeology and aesthetics. *World Archaeology* 33(2), 163–7.
Gosden, C. 2005. What do objects want? *Journal of Archaeological Method and Theory* 12(3), 193–211.
Grogan, E. 1991. Appendix: radiocarbon dates from Brugh na Bóinne. In G. Eogan, Prehistoric and early historic change in Brugh na Bóinne, 126–7. *Proceedings of the Royal Irish Academy* 91C, 105–32.
Hallowell, A. I. 1955. *Culture and Experience*. New York: Schocken Books.

Hallowell, A. I. 1975. Ojibwa ontology, behaviour and world view. In D. Tedlock and B. Tedlock (eds), *Teachings from the American Earth: Indian Religion and Philosophy*, 141–78. New York: Liveright.
Herity, M. 1974. *Irish Passage Graves: Neolithic Tomb-Builders in Ireland and Britain, 2500 BC*. Dublin: Irish University Press.
Hoffman, D. 2000. *Visual Intelligence: How We Create What We See*. New York: W. W. Norton.
Hollan, D. 2000. Constructivist models of the mind, contemporary psychoanalysis, and the development of culture theory. *American Anthropologist* 102, 538–50.
Ingold, T. 2000. *The Perception of the Environment: Essays in Livelihood, Dwelling and Skill*. London: Routledge.
Ingold, T. 2007. Materials against materiality. *Archaeological Dialogues* 14(1), 1–16.
James, W. 1995. *Selected Writings*. London: Orion.
Jones, A. 2004. By way of illustration: art, memory and materiality in the Irish Sea and beyond. In V. Cummings and C. Fowler (eds), *The Neolithic of the Irish Sea: Materiality and Traditions of Practice* 202–13. Oxford: Oxbow.
Jones, A. 2005. Lives in fragments? Personhood and the European Neolithic. *Journal of Social Archaeology* 5(2), 193–224.
Jones, A. and Bradley, R. 1999. The significance of colour in European archaeology. *Cambridge Archaeological Journal* 9(1), 112–14.
Jones, A. and MacGregor, G. 2002. Introduction: wonderful things – colour studies in archaeology from Munsell to materiality. In A. Jones and G. Macgregor (eds.), *Colouring the Past: The Significance of Colour in Archaeological Research*, 1–21. Oxford: Berg.
Kirk, T. 1993. Space, subjectivity, power and hegemony: megaliths and long mounds in Earlier Neolithic Brittany. In C. Tilley (ed.), *Interpretative Archaeology*, 181–223. Oxford: Berg.
Knappett, C. 2002. Photographs, skeuomorphs and marionettes: some thoughts on mind, agency and object. *Journal of Material Culture* 7(1), 97–117.
Knappett, C., 2006. Beyond skin: layering and networking in art and archaeology. *Cambridge Archaeological Journal* 16(2), 239–51.
Komlosy, A. 2004. Procession and water splashing: expressions of locality and nationality during Dai New Year in Xishuangbanna. *The Journal of the Royal Anthropological Institute* 10, 351–73.
Latour, B. 1993. *We Have Never Been Modern*. Trans. by C. Porter. Hertfordshire: Harvester Wheatsheaf.
Latour, B. 1999. *Pandora's Hope: Essays on the Reality of Science Studies*. Cambridge: Harvard University Press.
Latour, B. 2000. The Berlin key or how to do words with things. In P. M. Graves-Brown (ed.), *Matter, Materiality and Modern Culture*, 10–21. London: Routledge.
Layton, R. 1991. *The Anthropology of Art*. Cambridge: Cambridge University Press.
Lewis-Williams, J. D. and Dowson, T. A. 1990. Through the veil: San rock paintings and the rock face. *South African Archaeology Bulletin* 45, 5–16.
Lewis-Williams, J. D. and Pearce, D. 2005. *Inside the Neolithic Mind: Consciousness, Cosmos and the Realm of the Gods*. London: Thames and Hudson.
Lynch, F. 1970. *Prehistoric Anglesey: The Archaeology of the Island to the Roman Conquest*. Llangefni: The Anglesey Antiquarian Society.
Lynch, F. 1973. The use of the passage in certain passage graves as a means of communication rather than access. In G Daniel and P Kjærum (eds.), *Megalithic Graves and Ritual: Papers Presented at the III Atlantic colloquium, Moesgård 1969*, 147–61. Copenhagen: Jutland Archaeological Society.
Lynch, F. 1998. Colour in prehistoric architecture. In A. Gibson and D. D. A. Simpson (eds.), *Prehistoric Ritual and Religion: Essays in Honour of Aubrey Burl*, 62–7. Stroud: Sutton.
Macalister, R. A. S. 1939. *Newgrange, Co. Meath*. Dublin: Stationery Office

McMann, J. 1994. Forms of power: dimensions of an Irish megalithic landscape. *Antiquity* 68, 525–44.

Meighan, I., Simpson, D. and Hartwell, B. 2002. Newgrange – sourcing of its granitic cobbles. *Archaeology Ireland* 16(1), 32–35.

Meskell, L. 2005. Introduction: object orientations. In L. Meskell (ed.), *Archaeologies of Materiality*, 1–17. Oxford: Blackwell Publishing.

Mitchell, G. F. 1992. Notes on some non-local cobbles at the entrances to the passage-graves at Newgrange and Knowth, Co. Meath. *Journal of the Royal Society of Antiquaries of Ireland* 122, 128–45.

Mitchell, G. F. 1997. The geology of the Bend of the Boyne. In T. Condit and G. Cooney (eds.), *Brú na Bóinne*, 5–6. Dublin: Archaeology Ireland.

Mithen, S. 1996. *The Prehistory of the Mind: A Search for the Origins of Art, Science and Religion*. London: Thames and Hudson.

Mithen, S. 1999. Symbolism and the supernatural. In R. Dunbar, C. Knight and C. Power (eds.), *The Evolution of Culture: An Interdisciplinary View*, 147–69. Edinburgh: Edinburgh University Press.

Molloy, B. 2004. Unique prehistoric musical instrument discovered in County Wicklow. *Archaeology Ireland* 18(2), 5.

Ortiz, A. 1972. Ritual drama and the Pueblo world view. In A. Ortiz (ed.), *New Perspectives On the Pueblos*, 135–161. Albuquerque: University of New Mexico Press.

O'Kelly, C. 1971. *Illustrated Guide to Newgrange*. Wexford: John English and Co. Ltd.

O'Kelly, M. J. 1982. *Newgrange: Archaeology, Art and Legend*. London: Thames and Hudson.

O'Kelly, M. J. and O'Kelly, C. 1983. The tumulus of Dowth, Co. Meath. *Proceedings of the Royal Irish Academy* 83C, 135–190.

O'Sullivan, M. 1986. Approaches to passage tomb art. *Journal of the Royal Society of Antiquaries of Ireland* 116, 68–83.

O'Sullivan, M. 1993. *Megalithic Art in Ireland*. Dublin: Country House.

O'Sullivan, M. 1996. Megalithic art in Ireland and Brittany: divergence or convergence? *Revue Archéologique de l'Ouest*, supplément n° 8, 81–96.

O'Sullivan, M. 1997. On the meaning of megalithic art. *Brigantium* 10, 23–35.

O'Sullivan, M. 2005. *Duma na nGiall: The Mound of the Hostages, Tara*. Bray: Wordwell.

Ouzman, S. 2001. Seeing is deceiving: rock art and the non-visual. *World Archaeology* 33(2), 237–56.

Overing, J. and Passes, A. 2000. Introduction: conviviality and the opening up of Amazonian anthropology. In J. Overing and A. Passes (eds.), *The Anthropology of Love and Anger: The Aesthetics of Conviviality in Native Amazonia*, 1–30. London: Routledge.

Pearson, J. L. 2002. *Shamanism and the Ancient Mind: A Cognitive Approach to Archaeology*. New York: AltaMira Press.

Pearson, M. and Shanks, M. 2001. *Theatre/Archaeology*. London: Routledge.

Phelan, P. R. 1994. *The Magang Ceremony and Head-Hunting*. Kota Kinabalu: Department of Sabah Museum.

Powell, T. G. E. 1938. The passage graves of Ireland. *Proceedings of the Prehistoric Society* 4(2), 239–248.

Richards, C. 1993. Monumental choreography: architecture and spatial representations in Late Neolithic Orkney. In C. Tilley (ed.), *Interpretative Archaeology*, 143–78. Oxford: Berg.

Richards, C. 1996. Henges and water: towards an elemental understanding of monuments and landscape in Late Neolithic Britain. *Journal of Material Culture* 1(3), 313–36.

Ridington, R and Ridington, T. 1975. The inner eye of shamanism and totemism. In D. Tedlock and B. Tedlock (eds.), *Teachings from the American Earth: Indian Religion and Philosophy*, 190–204. New York: Liveright.

Rose, H. and Rose, S. (eds.) 2000. *Alas, Poor Darwin: Arguments Against Evolutionary Psychology*. London: Jonathan Cape.

Russell, I. 2004. Freud and Volkan: psychoanalysis, group identities and archaeology. *Antiquity* 80, 185–95.

Russell, I. 2007. Objects and agency: some obstacles and opportunities of modernity. *Journal of Iberian Archaeology* 9/10, 72–87.

Sahlins, M. D. 1977. *The Use and Abuse of Biology: An Anthropological Critique of Sociobiology*. London: Tavistock Publications.

Sanches, M. J., 2006. Passage-graves of northwestern Iberia: setting and movements. An approach to the relationship between architecture and iconography. *Journal of Iberian Archaeology* 8, 127–58.

Saunders, N. J. 2002. The colours of light: materiality and chromatic cultures of the Americas. In A. Jones and G. Macgregor (eds.), *Colouring the past: the significance of colour in archaeological research*, 209–226. Oxford: Berg.

Schefer, J-L. 1999. *Questions d'art Paléolithique*. Paris: P.O.L.

Seeger, A. 2002. Music and dance. In T. Ingold (ed.), *Companion encyclopedia of anthropology*, 686–705. London: Routledge.

Shee Twohig, E. 1973. Techniques of Irish passage grave art. In G Daniel and P Kjærum (eds.), *Megalithic Graves and Ritual: Papers Presented at the III Atlantic colloquium, Moesgård 1969*, 163–72. Copenhagen: Jutland Archaeological Society.

Shee Twohig, E. 1981. *The Megalithic Art of Western Europe*. Oxford: Clarendon Press.

Shee Twohig, E. 1990. *Irish Megalithic Tombs*. Princes Risborough: Shire Books.

Shee Twohig, E. 2000. Frameworks for the megalithic art of the Boyne valley. In A. Desmond, G. Johnstone, M. McCarthy, J. Sheehan and E. Shee Twohig (eds.), *New Agendas in Irish Prehistory: Papers in Commemoration of Liz Anderson*, 89–105. Bray: Wordwell.

Sheridan, J. A. 1985/1986. Megaliths and megalomania: an account, and interpretation, of the development of passage tombs in Ireland. *Journal of Irish Archaeology* 3, 17–30.

Sheridan, J. A. 2003. The chronology of Irish megalithic tombs. In G. Burenhult (ed.), *Stones and Bones: Formal Disposal of the Dead in Atlantic Europe During the Mesolithic-Neolithic interface 6000–3000 BC. Archaeological conference in honour of the late professor Michael J. O'Kelly*, 69–73. BAR International Series 1201.

Stam, R. 1989. *Subversive Pleasures: Bakhtin, Cultural Criticism and Film*. London: The John Hopkins University Press.

Stout, G. 2002. *Newgrange and the Bend of the Boyne*. Cork: Cork University Press.

Strathern, M., 1988. *The Gender of the Gift: Problems with Women and Problems with Society in Melanesia*. Berkeley (CA): University of California Press.

Squair, R. 1998. The Neolithic of the Western Isles. Unpublished PhD Thesis submitted to Glasgow University.

Taçon, P. S. C. 1999. All things bright and beautiful: the role and meaning of colour in human development. *Cambridge Archaeological Journal* 9(1), 120–23.

Thomas, J. 1990. Monuments from the inside: the case of the Irish megalithic tombs. *World Archaeology* 22, 168–178.

Tilley, C. 1996. *An Ethnography of the Neolithic: Early Prehistoric Societies in Southern Scandinavia*. Cambridge: Cambridge University Press.

Tilley, C., Hamilton, S., Harrison, S. and Anderson, E. 2000. Nature, culture, clitter: distinguishing between cultural and geomorphological landscapes; the case of hilltop tors in south-west England. *Journal of Material Culture* 5(2), 197–224.

Tringham, R. 2005. Weaving house life and death into places: a blueprint for a hypermedia narrative. In D. Bailey, A. Whittle and V. Cummings (eds), *(Un)settling the Neolithic*, 98–111. Oxford: Oxbow Books.

Turner, V. 1967. *The Forest of Symbols: Aspects of Ndembu ritual*. Ithaca: Cornell University Press.

van Gennep, A. 1960. *The Rites of Passage*. Trans. by M. B. Vizedom and G. L. Caffee. London: Routledge and Kegan Paul.

Vinnicombe, P. and Mowaljarlai, D. 1995. That rock is a cloud: concepts associated with rock images in the Kimberley region of Australia. In K. Helskog and B. Olsen (eds), *Perceiving Rock Art: Social and Political Perspectives*, 228–46. Oslo: Novus forlag.

Wade, N. 1990. *Visual Allusions: Pictures of Perception*. Hare: Erlbaum.

Wade, N. 1998. *A Natural History of Vision*. Cambridge: MIT Press.

Wade, N., and Swanston, M. 2001. *Visual Perception: An Introduction*. Hare: Psychology Press.

Wallis, R. J. 2003. *Shamans/neo-shamans: Contested Ecstasies, Alternative Archaeologies and Contemporary Pagans*. London: Routledge.

Watson, A. 2001. The sounds of transformation: acoustics, monuments and ritual in the British Neolithic. In N. S. Price (ed.), *The Archaeology of Shamanism*, 178–92. London: Routledge.

Watson, A. and Keating, D. 1999. Architecture and sound: an acoustic analysis of megalithic monuments in prehistoric Britain. *Antiquity* 73, 325–336.

Warner, W. L. 1958. *A Black Civilization: A Social Study of An Australian Tribe*. Chicago: Harper and Bros.

Whitley, D. S., Dorn, R. I., Simon, J. M., Rechtman, R. and Whitley, T. K. 1999. Sally's rockshelter and the archaeology of the vision quest. *Cambridge Archaeological Journal* 9(2), 221–247.

Wilde, W. 1849. *The Beauties of the Boyne and the Blackwater*. Reprinted [2003]. County Galway: Kevin Duffey.

Wilkins, A., Nimmo-Smith, I., Tait, A., McManus, C., Della Sala, S., Tilley, A., Arnold, K., Barrie, M. and Scott, S. 1984. A neurological basis for visual discomfort. *Brain* 107, 989–1017.

Witmore, C. 2006. Vision, media, noise and the percolation of time: symmetrical approaches to the mediation of the material world. *Journal of Material Culture* 11(3), 267–292.

Webmoor, T. and Witmore, C. forthcoming. 'What is gathered under the banner of the social? On the deprivations of dialectics in understanding humanity through a "social archaeology"'. *Norwegian Archaeology Review* 40(2).

Zeki, S. 1999. *Inner Vision: An Exploration of Art and the Brain*. Oxford: Oxford University Press.

Chapter 10
On Mediation and Material Agency in the Peircean Semeiotic

Christopher M. Watts

Abstract This chapter seeks to advance our understanding of material agency through an interpretive framework fashioned from the semeiotic ideology of Charles Sanders Peirce (1839–1914). In doing so, it attempts to move beyond a rote recital of Peirce's sign types and their lineaments and toward a larger reading of his philosophical outputs, examining potential points of contact between material agency and Peirce's thinking on semeiotic functioning. Owing to the contours of a creative mind steeped in mathematics and logic, his is a canon marked by heroic theorising, labyrinthine reasoning and runaway terminology. As such, uncharitable interpretations of Peirce's writing often evoke words such as 'impenetrable' or 'torturous,' but it is nevertheless a literature that commands our attention, chiefly because of its non-anthropocentric, anti-Cartesian emphasis on semeiotic mediation. In the Peircean framework, semeiosis is neither bound up in language nor contingent on human consciousness, but rather exists as a relative and relational property tethered to particular experiential settings. Where the human subject is implicated, perception, cognition and belief were understood by Peirce to be engendered by a sensory experience of signs. The phenomenological underpinnings of these themes are explored throughout this chapter and illustrated with reference to a case study involving precontact Aboriginal pottery from southwestern Ontario, Canada.

Introduction

> [D]ualism in its broadest legitimate meaning [is] the philosophy which performs its analyses with an axe, leaving as the ultimate elements, unrelated chunks of being... (Peirce, 1958–1965, Collected Papers [hereafter CP[1]] 7.570).

C.M. Watts
Department of Anthropology, University of Toronto, Toronto, Ontario, M5S 2S2, Canada
e-mail: christopher.watts@utoronto.ca

[1] I follow the convention of quoting Peirce with reference to a standardised notation consisting of two initials, which refers to the source of the material, followed by numbers denoting

C. Knappett, L. Malafouris (eds.), *Material Agency*,

DOI: 10.1007/978-0-387-74711-8_10, © Springer Science+Business Media, LLC 2008

In what may be described a groundswell, as other chapters in this volume attest, scholarly interest in non-anthropogenic forms of agency has been thrust to the forefront of concern in a variety of disciplines. Among archaeologists, such a harbinger may be seen as particularly welcome, even if the impassioned pleas by Dobres and Robb (e.g., 2000) to define what exactly is meant by agency have yet to be satisfactorily addressed. Indeed, it might be best to regard such a state of affairs as fortuitous, particularly when considered in light of recent calls for a counter-modern approach to archaeology (e.g., Thomas, 2004, 2007; Schnapp et al., 2004) and the implications these have for sacrosanct ontological dualisms involving people and things. Mercifully, where definitions of agency are concerned, an entrenched doctrinal frame does not yet exist, for I suspect if it did, it would now need to specify how non-humans can also be said to 'act'.

One model which speaks to this issue, and which has been gaining traction among archaeologists in recent years (e.g., Boast, 1997:187; Knappett, this volume; Olsen, 2003; Watts, 2007, 2008; Whitridge, 2004) is Actor Network Theory or ANT. As both a conceptual device and a methodological tool, ANT adopts as its defining criterion the notion that both humans and non-humans serve as co-constituents within vast heterogeneous spheres of influence. It is possible to see within ANT and its earlier incarnation, Science and Technology Studies (STS), a distinct lineage of anti-essentialism that seeks to conflate many of the dualisms found in modern scientific thought, including fundamental oppositions between nature/culture and subject/object. It has been suggested (Latour, 1993:10–12) that by subscribing to such distinctions, we habitually ignore acts of mediation—how admixtures or a *sui genera* of nature/culture hybrids ultimately emerge through our interactions with things. Along these lines, agency in ANT is seen not as a fixed and unchanging property of the active human subject but rather as something engendered by the interdependence that exists between agential entities, both human and non-human alike, and exercised through their linkages.

ANT scholars such as Bruno Latour, Michel Callon, John Law and Madeleine Akrich have devoted considerable attention to this notion of human/non-human interdependence and have attempted to further the idea in a variety of innovative ways. One such approach involves what John Law (1999:4; see also Akrich and Latour, 1992) refers to as a "semiotics of materiality" wherein both human and non-human 'actants' are brought forward through relationality, that is, by virtue of their respective positions (as nodes) within networks of meaning. Much of this thinking is underpinned by the work of structuralist semiotician Algirdas Greimas (e.g., 1966) who attempted to illuminate the narrative structure of texts. As discussed by Hawkes (1977:87–95; see also Lenoir, 1994:125; Whitridge, 2004:454), Greimas held that a limited number of elemental binary oppositions

volume and paragraph (where applicable). For example, 'CP 2.247' refers to Volume II, paragraph 247 of *The Collected Papers of Charles S. Peirce* (1958–1965). Passages in the *Collected Papers*, as well as those from other primary sources, are typically cited without the titles of the individual manuscripts, although Peirce scholars sometimes cite approximate dates of composition (not adopted herein).

(i.e., subject/object, sender/receiver, helper/opponent) formed the basis of a universal *modèle actantiel* or 'actantial model' from whose structure the lineaments of individual narratives emerge. Actants in the Greimas model could be either human or non-human, for example, a protagonist or a holy grail, but they are always functionally defined by the positions they occupy in broader narrative journeys, just as they are, syntactically, in language.

ANT is principally occupied with this notion of relationality among actants and follows the semiology of Ferdinand de Saussure who suggested that a deeply seated *langue* (competence) gives rise to specific instantiations of *parole* (performance). Saussure believed that it was a structured set of differences which served to codify a word (signifier) and its meaning (signified) as part of an arbitrary, dyadic entity known as the linguistic sign. But as Hodder (e.g., 1989:257, 1992:201–202) notes, it is impossible to address signification in material culture using the Saussurean approach. In many cases, signification systems among non-humans do not turn on convention, as is the case with language, but rather on a varied set of articulations that inhere between sign and signified and which give rise to more complex processes of semiotic mediation. Here I would like to suggest that the veracity of ANT and indeed of any model which seeks to understand how non-humans might 'act', can be advanced through a consideration of the semeiotic[2] of Charles Sanders Peirce (1839–1914) including its basis in synechism, its three-fold categories of being and finally, its triadic relations of performance, comparison and thought. Taken together, these facets of Peirce's thinking can help us not only to overcome the deeply seated dualisms noted above, but also to work toward a conception of material agency as defined by semeiotic principles. After elaborating upon these two themes, I describe a recent approach to classifying and interpreting precontact Aboriginal pottery production from southwestern Ontario, Canada, within the Peircean scheme and conclude by considering more generally how such a framework allows us to speak to signification in material culture.

On Signification in the Peircean Semeiotic

The work of Charles Sanders Peirce, while largely unpublished and unrecognised during his lifetime, has been the subject of renewed interest in a multitude of disciplines. Nowhere is this more apparent than in the fields of anthropology and archaeology, as suggested by a review of recent literature (e.g., Bauer, 2002; Capone and Preucel, 2002; Coben, 2006; Gardin, 1992; Gell, 1998; Keane,

---

[2] In keeping with a common practice, I use the terms *semeiotic* and *semeiosis* in this chapter to refer specifically to Peirce's triadic theory of the sign and conception of semeiotic mediation respectively. The term *semiotic* is used elsewhere in this chapter to denote the more general study of signs, while the term *semiology* is deployed only in connection with the brand of semiotics developed by Ferdinand de Saussure.

2003a,b, 2005; Knappett, 2002, 2005; Lele, 2006; Parmentier, 1997; Preucel and Bauer, 2001; Preucel, 2006). Peirce's voluminous writings are mined by scholars to a variety of ends, in large part because he contributed to a broad range of epistemological and ontological debates. He was the founder of pragmatism—the notion that a theory or proposition may be judged by its observable practical consequences—and a pioneer in the study of perception and thought, both of which he believed could be examined through a formal doctrine of signs called *semeiotic*. Such an inquiry was understood by Peirce to be rooted in the continuities that come about between internal representation and external reality (CP 1.551) including the extent to which cognition proceeds from an articulation of signs native to both realms. In this way, Peirce intended his semeiotic to be based on a brand of phenomenology he also developed known as *synechism*, from the Greek *synechismos*, wherein "all that exists is continuous" (CP 1.172). A significant anti-Cartesian ramification of this view, to use Peirce's words, is that "consciousness has a bodily and social dimension, the latter originating outside the individual self" (CP 7.575). As subject and object conflate in the process of signification, he believed, interpretation could not be seen as something projected onto the world by an inviolate preserve of the thinking self. Instead, interpretations proceed from signs, or rather *are* signs and it is this that renders the Peircean semeiotic in one way distinct from Saussurean semiology. While Saussure regarded the sign as a wholly abstract and thoroughly *dyadic* entity employed by members of a speech community, Peirce (CP 1.480) regarded it as a relatum which mediates between an object (that which the sign stands for) and an interpretant (the referent of the semeiotic process) as part of a single *triadic* relation. Importantly, Peirce also acknowledged that the sign was more than simply a vehicle in that it mediated, through what he called the "ground" (CP 2.228), the character of representation itself.

Equally important, however, is Peirce's notion of the interpretant, which is not a person but rather the "proper significate effect" (CP 5.475) or result of sign functioning. The interpretant functions to approximate the object-sign relationship through a sequent representation which is informed by the object and directly brought about by the sign (CP 6.347; see also CP 5.484). For Peirce, then, qualia are inextricably conditioned by semeiosis; the world and its products are disclosed by interpretants. But disclosed to who (or what)? Although Peirce often wrote that the interpretant is sited in "the person," he once noted in a 1908 letter to Victoria, Lady Welby (1837–1912) that such a pre-condition was "a sop to Cerberus, because I despair of making my own broader conception understood" (Peirce, 1977, Semiotics and Significs [SS] 80–81). Indeed, Peirce would sometimes use the term "quasi-mind" to describe the locus of the interpretant, itself a sign, and omit any reference to human beings as the sole category of interpreters, as in the following passage:

> ... a Sign has an Object and an Interpretant, the latter being that which the Sign produces in the Quasi-mind that is the Interpreter by determining the latter to a feeling, to an exertion, or to a Sign, which determination is the Interpretant [CP 4.536].

Accordingly, Peirce's semeiotic requires only that an interpretant be formed—the significate effect of sign functioning and this in no way presupposes human consciousness. Hence, another major distinction between the thinking of Saussure and Peirce and of particular import in light of this volume and its aims is that the interpretant is neither anthropocentrically nor cognitively inclined. "Thought," to Peirce, "is not necessarily connected with a brain. It appears in the work of bees, crystals and throughout the purely physical world; and one can no more deny that it is really there, than that the colours, the shapes, etc., of objects are really there" [CP 4.551]. Not only, therefore, is all thought in signs, but signs are inherently extra-human.

That Peirce regarded the sign-object relationship as productive of an autonomous interpretant is a central theme of his semeiotic and perhaps the most accessible point of entry through which to explore how humans and non-humans commingle in semeiosis. But before doing this, it is instructive to first consider the object-sign-interpretant model as a manifestation of Peirce's broader tripartite doctrine of categories. As Savan (1989:15) notes, although the categories play a significant role in virtually all aspects of Peirce's thinking, they are absolutely vital to an understanding of his semeiotic.

Like Kant and Hegel before him, Peirce devoted considerable attention throughout his career to developing a phenomenological scheme which could capture and classify irreducibly all that comes before the mind. Peirce referred to this as the *phaneron*, or the sum total of everything "present at any time to the mind in any way" (CP 1.186). He set out to show that whatever phenomena may be experienced are at once comprised of *Firsts, Seconds and Thirds*—states of being which articulate with one another to form "the triune Reality" (CP 5.431). What is more, Peirce argued that each of these states could be further defined as having either a monadic, dyadic or triadic form of relationality. Put briefly, a First may be defined as a purely qualitative feeling or possibility, such as the colour red or a musical note, which is not embodied and therefore can only be grasped through abstraction. The sign relatum finds its proper domain in Firstness. As a state, Firstness is defined by monadic relations: feelings or possibilities exist in the world, but only to themselves and not to things or events—"it is the mode of being which consists in its subject's being positively such as it is regardless of aught else" (CP 1.25). To this, Peirce adds that a First "cannot be articulately thought: assert it, and it has already lost its characteristic innocence; for assertion always implies a denial of something else. Stop to think of it, and it has flown" [CP 1.357].

By contrast, we react to Seconds as unmediated 'brute facts,' which are embodied as objects of reality and operate in a dyadic mode of relationality. Secondness reveals essential dualisms (e.g., action/reaction, cause/effect, etc.) and renders capable our understanding of things in themselves "in the sense of being present regardless of the perceiver's will or wish" (CP 5.462). The object relatum belongs in this category and by virtue of this is able to direct and limit sign action by "forcing its way to recognition as something other than the mind's creation" (CP 1.325). True semeiotic mediation, however, does not take place without *Thirds* which ultimately allow interpretants to transcend an external reality of brute facts and dynamic reactions. By necessity, interpretants are triadic in nature since they mediate between sign and object, possibilities (Firsts) and actualities (Seconds). Reason, understanding, thought and purpose can only take place within the triadic mode of Thirdness, which gravitates toward the creation of rules or laws that make the possible 'actual'.

In defining Secondness, Peirce rebuffed Hegel's brand of idealist phenomenology for failing to take seriously the "outward clash" of experience (CP 8.41). To Peirce, it is the outward clash that links perception (First), volition (Second) and cognition (Third) through an embodied and therefore sensuous character of experience. In doing so, it brings into relation Seconds such as the credible and the fallible, the real and the interpreted and the particular and the general (see Hookway, 1985). Something is only fully present as an interpretant through its external manifestation—"[w]hat passes within we only know as it is mirrored in external objects" (CP 8.144) and thus it cannot be known through internal reflection alone. As is discussed in some detail below, the sign type known as indices permits these connections to be made.

Set against the backdrop of Peirce's categories, we can see more fully how the nature of semeiosis is intertwined with a triadic conception of phenomena. The conjoined nature of the three states (First, Second and Third) serves as a platform for the corresponding trigeminal operation of semeiosis (sign, object, interpretant). As part of Peirce's phenomenology, Firsts cannot be verily separated from Seconds: colour, for example, may be *abstracted* (Peirce uses the word 'prescinded') but not *realized* outside of the space which it occupies, while neither can be extracted from the Third through which the dialogue of experience passes. Similarly, objects and their signs are 'grounded' in a dyadic relationship—"the entire universe... is perfused with signs, if it is not composed exclusively of signs" (CP 5.448n)—but it is the interpretant that designates the manner in which these signs are revealed. One of the more significant contributions such a perspective makes to the topic at hand, however, issues from the implication that human experience in the Peircean scheme is not an immanent undertaking. It is given by the irreducible unity of synechism, the congruence of both humans and non-humans as loci for sign activity and the interactions that take place between them. From this, we might profitably consider the semeiotic as imbued with agency, inasmuch as sign functioning to Peirce was seen as a stepwise yet subjective product of the continuities found between objects of reality and representation. It becomes possible in such a light to see material culture as more than just a *tabula rasa* etched with the accounts of a Cartesian consciousness and filled with 'symbolic' content. Moreover, we can look to things as generative of protean effects without deferring to dialectics between structure and agency or metanarratives of meaning, as in the Greimas scheme. Semeiotic provides a way to recast the character of experience through a more consonant, sign-based phenomenology and it is with this in mind that I intend to frame my review of Peirce's sign trichotomies (relations) below.

The Sign Trichotomies

Peirce developed the object-sign-interpretant relata in keeping with his categories of Firstness, Secondness and Thirdness but he also regarded sign relations as being equally caught up in such an arrangement. Hence, every sign in the Peircean scheme has three states of being, each of which corresponds with three stages or moments of semeiotic functioning. Roughly, these three stages specify the nature of (1) the sign in relation to its object, (2) the sign in relation to itself and (3) the sign in relation to its interpretant, that is, how it may be interpreted to represent an object. The articulations between sign relations and their products are presented in Table 10.1. Importantly, it should be noted here that the categories of Firstness, Secondness and Thirdness interact in varied ways with the relations of performance, comparison and thought such that Table 10.1 can be regarded as a paradigmatic sign-generating classification. Accordingly, this scheme offers up 27 possible sign combinations, although 17 combinations may be excluded from further consideration by logical necessity. This point, along with the nature of each trichotomy, is discussed in more detail below.

Peirce is perhaps best known for his "triadic relations of performance" (CP 2.234; 2.247–249) which specify the kind of ground that exists between an object and its sign. The first of these forms, as a First, is known as an *icon*, and connects object to sign through mimesis or formal resemblance (e.g., as seen in a photograph). The second form, known as an *index*, can be characterised by a deictic or existential connection between object and sign, such as in the relationships between smoke and fire, weathervane and wind, or cough and cold. In these cases, as a Second, the index points to or brings about the interpretant through causality rather than likeness. The final sign form, referred to as a *symbol*, is considered a Third since it relates object to interpretant by means of a conventional association. Words in a language, for example, are symbols, and this form closely approximates the Saussurean definition of a linguistic sign.

Crucially, Peirce regarded the triadic relations of performance as a contiguous set replete with an internal logic resistant to circumscription. As such, icon, index and symbol cannot be seized upon as mutually exclusive types of

Table 10.1 Peirce's three trichotomies

	Triadic relations of performance (signs in relation to their objects)	Triadic relations of comparison (signs in relation to themselves)	Triadic relations of thought (signs in relation to their interpretants)
First	Icon	Qualisign	Rheme
Second	Index	Sinsign	Dicent
Third	Symbol	Legisign	Argument

signs but rather occur as processual moments in a graduated continuum of semeiotic functioning. As Peirce (CP 2.247) notes, an icon is a sign by virtue of possessing in itself certain qualities which it would also possess if the interpretant and the object did not exist. Thus, while an icon could not function as a sign without its object, as a sign its attributes exist independently of both object and interpretant (i.e., in a monadic relationship). The qualities inherent in an index, on the other hand, can also be seen to exist apart from the interpretant but not the object, to which it is dyadically tethered in both space and time. A symbol, meanwhile, could not exist without both object and interpretant and consequently is the only sign type that is predicated by a triadic relationship (see Parmentier, 1994:6–7; Short, 2004:223). This notion may be more fruitfully examined with an example.

If we are outside flying a kite, we may observe its movements and understand them in connection with the velocity and direction of the wind. When such a connection is made, that is, the sign is recognised and understood *as* a sign (CP 2.231), then the ground between sign (kite direction) and object (wind direction) can be interpreted as an icon due to formal resemblance. The kite, however, could not function as an iconic sign without the phenomenon of the wind (its object), but as an icon it displays the properties it does without recourse to either object or interpretant. Nor could it function as an indexical sign, were there not a causal connection between object and sign, wind and kite, in that particular place and at that particular time. But the wind would still affect the kite regardless of whether or not we make such connections and it is in this sense that both icon and index can be seen to signify apart from an interpretant. To understand the ground between object and sign as a symbol, by using the English word 'wind' to describe the object for example, would *necessarily* entail an interpretant since this association is entirely motivated by an arbitrary (conventional) relationship.

To my mind (no pun intended), Peirce's definition of sign performance provides us with a critical first step toward realiing how material agency might operate. Through its finer grain and hierarchical scale, the object-sign-interpretant triad permits passage around the barrier to semiotic functioning encountered by Hodder (1992) and others (e.g., Shanks and Tilley, 1987); signification in material culture need not be reconciled with Saussurean notions of the epiphenomenal linguistic sign in order to proceed. Indeed, when Peirce writes that a sunflower, by turning toward the sun, "becomes by that very act fully capable, without further condition, of reproducing a sunflower which turns in precisely corresponding ways toward the sun and of doing so with the same reproductive power" can be considered a sign of the sun (CP 2.274), we see that a sign is distinguished by the contours of its significate effect rather than by any reference to human cognition. In the Peircean scheme, all signs display an agency of sorts, and this can be further explored with reference to the trichotomy of performance. Although icons and indices play an essential role in cognition, when they are recognised as such, they are neither defined by nor disposed toward perception and therefore can operate beyond human purview.[3] This much was observed by Gell (1998; see also Gosden, 2005) in his approach to art objects[4] as well as in his conception of the "inter-artifactual domain." But what is truly insightful about Peirce's thinking here is not that 'natural' signs exist (Aristotle referred to these as *semeions* in *On Interpretation*) but rather that such signs can be placed under the same analytical lens as those entirely motivated by 'convention'. That we can regard such natural signs as part of a continuous program of semeiotic functioning more than hints at the fallacy associated with dualisms such as nature/culture and subject/object. More to the point, though, it further suggests that what we would traditionally classify as 'symbolic', should now be seen as but one of three modalities whereby interpretants are mediated by object-sign articulations. To be sure, symbols evince a 'higher-order' or more complex degree of semeiotic mediation because they are thoroughly bound up in a conventional relationship, but they nonetheless incorporate indices to point to objects of signification, while indices require icons to make evident the substantive character of objects (CP 2.247–249; see also Parmentier, 1994:6–7). In other words, it is impossible in such a frame to regard the Cartesian *cogito*—that bastion of cognitive faculties reserved for the active human subject—as insulated from the effects of a semeiotic mediation so clearly coloured by the vagaries of experiential settings.

But the fact that the symbol is rooted in a conventional rather than contextual relationship between ground and interpretant also indicates that it can outlive the more immediate, experiential conditions which give rise to both iconicity and indexicality. This, of course, is readily apparent from our use of language, which instantiates symbols independently of the particular phenomena with which they are associated. Peirce was well aware of this distinction and sought to expand upon the logical possibilities of his sign types through a second trichotomy known as the "triadic relations of comparison" (CP 2.243–246). This trichotomy focuses on how the sign may be defined not in relation to its object but rather in relation to itself. The first of these forms, which corresponds with Firstness and iconicity, is known as the *qualisign* and refers to the 'quality' or qualitative possibility of the sign, for example, the colour green or a musical note, once it is embodied. The second, as a Second, is known as a *sinsign* or a 'singular' physical object or event. Sinsigns manifest themselves as indices, for example, a road sign, and incorporate qualisigns in much the same way as indices involve icons. The third sign form, typified by symbols and Thirdness, is called a *legisign* and is defined by 'artificial' laws or rules which govern how the sign operates. Language, by working through an abstract yet principled set of

[3] In a similar vein, Thomas Sebeok (e.g., Sebeok and Umiker-Sebeok, 1992; Sebeok et al., 1999) has long championed the application of Peircean semeiotic principles to the study of sign use among animals (*zoosemiotics*) and, more generally, to a study of sign processes among all living things (*biosemiotics*).

[4] It should be noted, following Preucel (2006:265n), that Gell's treatment of indices, as well as the Peircean notion of abduction—"the process of forming an explanatory hypothesis" (CP 5.171) in response to a phenomenon of interest—were coloured by Eco (1976).

object-sign relationships, comprises a recurring set of legisigns, but typically any sign that is rule-bound, for example, the use of a whistle at a sporting event, may be classified as a legisign.

Strictly speaking, however, because legisigns are made up of sinsigns, which in turn embody qualisigns, the context-specific occurrence of a legisign is referred to by Peirce using a special category of sinsigns known as *replicas*. A replica is not totally unlike what we would call an artifact 'type' in more traditional classificatory schemes. Indeed, Peirce introduces this term, along with 'token' (CP 4.537, 4.544) to describe a replica. To continue with this analogy, we classify materials in a typology by recognising and configuring attributes into rule-bound categories or legisigns—often with sobriquets such as Scottsbluff Point or Grooved Ware—the physical manifestations of which are replicas. What separates replica from sinsign here is the same as what separates convention from context: the replica is recognised as significant only by a community of users familiar with the category and the broader typology to which it belongs. Examined by the non-specialist, the replica in this case would not function as an instance of a legisign since this person lacks a requisite familiarity with the overarching law or rule by which a conventional rather than contextual relationship is specified.

To complicate matters, however, in further defining the triadic relations of performance, Peirce notes that while all symbols may be seen as legisigns, not all legisigns are symbols (CP 2.246). Parmentier (1994:9) illustrates this nicely using the second-person (singular) personal pronoun "you." By all accounts, this is a legisign since English speakers recognise the same word in all its various contextual applications, but as a legisign it represents its object by virtue of a ground not typically associated with symbols. In this case, "you" refers to a dynamical addressee (object) who nonetheless must be an integral co-constituent in every successful referential act involving the replica of "you." In other words, there is an indexical dimension, as in any demonstrative pronoun, implicit in the transitory nature of this word since the object to which the sign points hinges on a very specific rather than general occurrence (CP 2.259). Accordingly, Peirce specifies that indexical legisigns do not "possess the generality of purely conventional signs" (Peirce, 1967, Manuscripts in the Houghton Library of Harvard University [MS] 748).

This subtlety, in my opinion, also has important implications for material agency insofar as it argues for a definition of legisigns that is not wholly based on imputed laws. It advances the claims of Gell (1998) and Gosden (2005), for example, by illustrating how indexical legisigns can reference and subsequently engender (as interpretants) a subset of objects defined by materiality as opposed to an externalised symbolic content. As Parmentier (1997:50–51) notes, "[p]ottery style rarely functions symbolically; rather, style is an indexical legisign embodying an iconic legisign, and a particular pot in that style is an indexical sinsign—a 'replica,' in fact, since it is generated from a template which it (trivially) indexes." Consequently, where both analyst and artisan are concerned, we need to remember that the object is at once qualisign, sinsign and legisign when it is referenced as part of a triadic relationship. The hierarchy upon which Peirce bases this idea precludes a consideration of material culture as shaped exclusively by the 'social' since materiality entails qualities (qualisigns) embodied by individual things (sinsigns) which appear as representations or interpretants (in legisigns). Where things are involved, legisigns, in particular indexical legisigns, do not exist independently of the 'lower-order' functioning of monadic and dyadic relationships encapsulated by their presence. Perhaps more importantly, however, these properties, be they lustre, weight or hardness, need not await the arrival of an assigned symbolic interpretant in order to function as signs, though they will if specified as such, but rather 'act' or perform in ways that can potentially contribute to other sign relationships. When a replica exists as an instantiation of a legisign, in this case a pot of a particular style recognised as such by the analyst or artisan, the same replica also embodies qualities which have nothing to do with its status as a replica in this particular legisign.[5]

This will perhaps become clearer when we consider the third and final trichotomy, recognised by Peirce as the 'triadic relations of thought' (CP 2.250–253). In this arrangement, Peirce classifies the sign according to the way in which it presents its object to the interpretant. As a First, a *rheme* exhibits the mere possibility of an object and correlates with iconicity and qualisigns in the other trichotomies. A rheme is any sign which is neither true nor false and Peirce regarded it as analogous to a logical proposition wherein certain parts are blank. Until such time as these parts are filled in with values, the veracity of the proposition cannot be tested (see CP 4.560). A *dicent* (or dicisign) articulates with the domains of Secondness, indexicality and sinsignhood, and conveys information about its object as a 'brute fact'. Finally, the object of an *argument* appears to the interpretant as a law, and occupies the same analytical realm as a Third, symbol and legisign—that is, it will by necessity appear as part of a fully-formed triadic relationship.

As was noted in Table 10.1, if signs are examined at once in terms of how they relate to their objects (performance), themselves (comparison) and their interpretants (thought), the nine sign modalities will produce 27 possible combinations. Not all of these combinations are, however, given by logical necessity: symbols, for example, must be legisigns which excludes their articulation with either qualisigns or sinsigns. Similarly, as mere possibilities, Firsts cannot display any inherent complexity, which eliminates both indices and symbols from this state of being. Peirce also believed that interpretants could not represent the ground between object and sign as more complex than it actually is, meaning icons could not appear as indices or symbols while indices could not appear as symbols. Where icons are referenced, this excludes their being recognised as either dicents or arguments; for indices, this prevents their being considered as arguments. If we remove these combinations from further analysis, we are left with the ten combinations that appear in Table 10.2. In the same vein, logic

5 Arguably, there is a parallel here with Costall's attempts (e.g., 1995, 1997) to 'socialise' the Gibsonian notion of affordance, in that such qualities could be seen as the non-canonical affordances of the material (see also Knappett, 2005:47–49)

Table 10.2 The ten sign possibilities in the Peircean semeiotic (CP 2.264)

No.	Category	Name	Example
1	First	(Rhematic Iconic) Qualisign	a feeling of 'red'
2	Second	(Rhematic) Iconic Sinsign	an individual diagram
3	Second	Rhematic Indexical Sinsign	a spontaneous cry
4	Second	Dicent (Indexical) Sinsign	a weathervane, a photograph
5	Third	(Rhematic) Iconic Legisign	a diagram
6	Third	Rhematic Indexical Legisign	a demonstrative pronoun
7	Third	Dicent Indexical Legisign	a street cry
8	Third	Rhematic Symbol (-ic Legisign)	a common noun
9	Third	Dicent Symbol (-ic Legisign)	a proposition
10	Third	Argument (-I've Symbolic Legisign)	a syllogism

dictates that certain modes will produce redundancy (expressed in parentheses)—a qualisign, for instance, can only be expressed as an icon, and by virtue of this must be a rheme. Similarly, an argument can only be supported by a symbol which in turn entails a legisign.

Although one might think that this heuristic exhausts all possible forms of semeiotic functioning, Peirce would ultimately go on to propose ten trichotomies, based on the combinations presented in Table 10.2, yielding 66 classes of signs. But he also considered the extent to which each of these ten trichotomies was mutually exclusive, and thus ultimately divisible into three additional trichotomies, which if true would render 3^{10} or 59,049 sign classes! Peirce would never develop these classes and indeed regarded such a pursuit as prohibitive—"I have 3^{10} or 59049, difficult questions to carefully consider, and therefore I will not undertake to carry my systematical division any further, but will leave that for future explorers" [CP 8.343].

Not being one of these explorers, I only wish to point out here that Peirce's sign possibilities were intended to delineate a truly general semeiotic, one which could succeed in expressing the myriad phenomena of experience. In this sense, then, perhaps the more important upshots of Peirce's thinking are not to be found in his relentless ramifying but rather in the fecundity of semeiotic and the ways in which it takes up the inbuilt vitality and dynamism of signs. Although it is clear Peirce struggled to harmonise empirical observation with metaphysical being, in the same way he strained to marry ontic classification with ontological ineffability, his semeiotic nonetheless opens up and operationalises a unique inquiry into how the world exists. It can be seen to de-centre traditional views of the human subject as omnipotent while it dispenses with the tyrannies of an Aristotlean substance ontology and Cartesian rational consciousness. The world is not contained within a thing any more than it is held in some rarefied mind: it is brought forth through the triadic nexus of semeiosis. As signification that cannot be severed from its experiential settings, material agency is likewise enmeshed in semeiotic functioning. Things 'act' by delimiting and determining sign functioning, as discussed in connection with Peirce's triadic relations of performance. At the other end of the spectrum, the interpretant or 'proper significate effect' is mediately determined by such things, as well as an active influence on the interpreter, but is by no means reliant upon human consciousness. We enter into this relational nexus, however, when such signs come before the senses. Consequently, we find meaning in the world rather than fashion it, and Peirce's synechistic account of reality, coupled with his categories and trichotomies, paints such a dialogic portrait of experience and sign activity.

Illuminating Semeiotic Mediation in Late Woodland Pottery

Due to its extraordinary terminology and internal complexity, the Peircean semeiotic is an admittedly unwieldy device. It does, however, provide a useful means by which to consider how things 'act' and to think about the signification process as a succession of capable, contextual and conventional states of being. In doing so, it provides a unique way to locate and define the manifold where agential forces of materiality and humanity come together. I have found the above approach to be vital to my own understanding of precontact Aboriginal pottery production in southwestern Ontario, Canada, where as part of my doctoral research (Watts, 2007, 2008), I analysed over 800 earthenware vessels from seven Late Woodland Period (ca. AD 900–1300) sites. In this region, the Late Woodland is generally thought to comprise two adjacent cultural sequences known as the Western Basin and Ontario Iroquoian Traditions (Fig. 10.1). Western Basin groups likely consisted of mobile hunter-gatherers who aggregated in warm weather lacustrine and riverine environments followed by a pattern of inland dispersal during cold-weather months. Food production, specifically maize horticulture, is not a salient feature of Western Basin economies during this time; sites typically lack palisading and consist of one or two dwellings with an abundance, on warm-weather sites, of large storage pits. By comparison, between the tenth and fourteenth centuries AD, Ontario Iroquoian groups came to occupy semi-sedentary palisaded villages made up of four to five longhouses with large interior and exterior storage and refuse pit features.

During the time period in question, Late Woodland pottery generally consists of grit-tempered earthenware vessels, approximately four to 15 litres in volume, with vertical to everted rim orientations, collared or uncollared rim surfaces, constricted necks and pronounced shoulders. Vessels are typically decorated on interior, lip and exterior surfaces (above the shoulder) although undecorated specimens are also found, particularly within Iroquoian assemblages. Surface treatments primarily consist of cord roughening, which may be smoothed-over on rim and neck surfaces prior to the application of decoration. Where present, decoration consists of stamped and incised surface modifications which are organised into a series of horizontal bands. These modifications commonly feature discontinuous line or string elements (e.g., punctates) which, primarily among Western Basin groups, may be combined into a complex series of triangular- or diamond-shaped zones on vessel neck surfaces (Fig. 10.2).

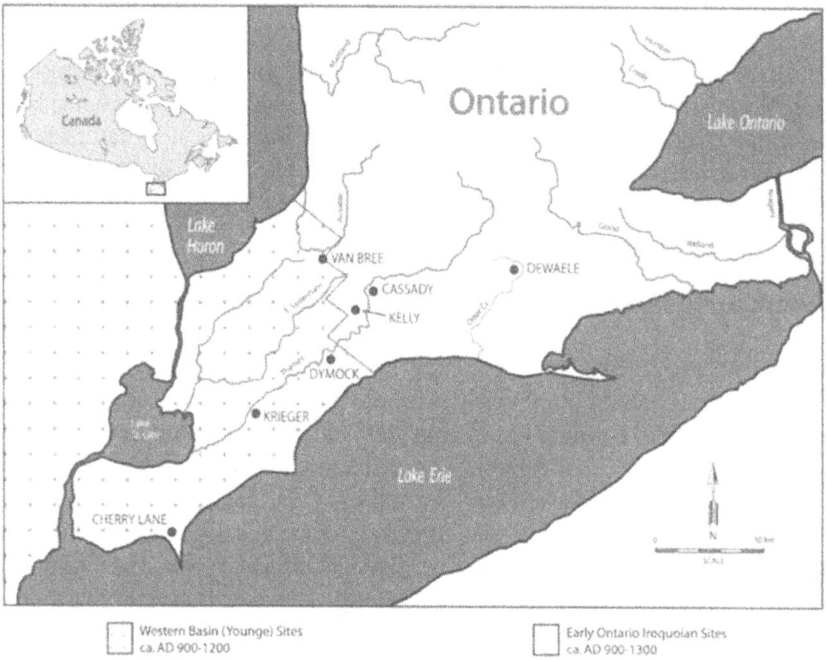

Fig. 10.1 Late Woodland Period traditions and sites examined in Watts (2008)

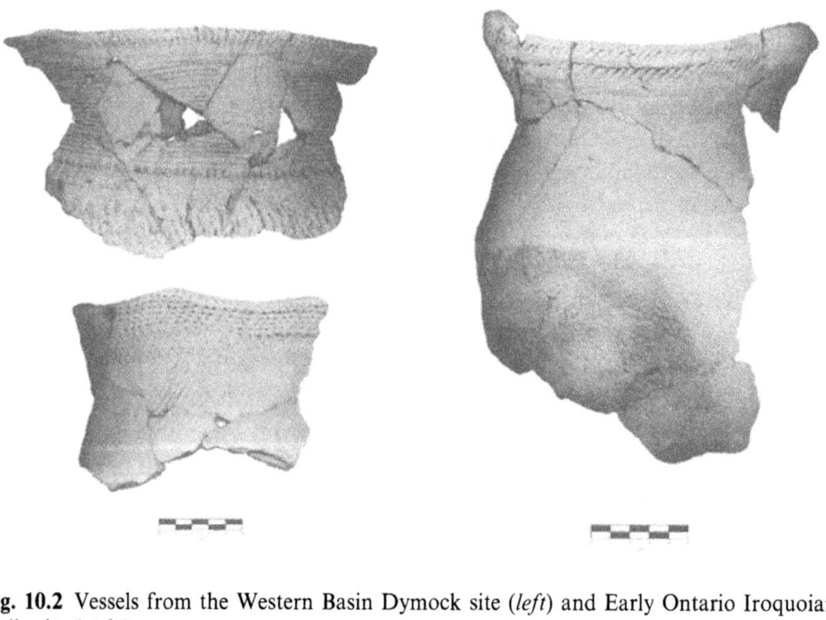

Fig. 10.2 Vessels from the Western Basin Dymock site (*left*) and Early Ontario Iroquoian Kelly site (*right*)

Rather than approach pottery form and decoration in terms of functional constraints and 'symbolic' behaviour, the latter of which might be treated as a concealed discourse impervious to interpretation, my analyses proceeded as follows. I worked under the assumption that qualities associated with potting clays (e.g., temper, colour, plasticity, organic content, etc.), along with the spatial configuration of morphological and decorative elements, would have been apprehended somatically through experience and embodied by vessel designs as Seconds. By virtue of their placement and use within varied contextual settings, I also began to consider how interpretants as Thirds might be differentially produced by the ground between sign and object in Iroquoian and Western Basin actor-networks. An intricate attribute-based coding scheme was used to classify both discrete and continuous properties of form (e.g., rim height, thickness, orientation, etc.) and decoration (e.g., tool, technique and motif use).

With regard to decorative attributes, these were further classified according to the symmetrical classes by which elements were organised and manipulated (Fig. 10.3). Analyses of symmetry typically proceed from the identification of a basic unit or "fundamental part" (Shepard, 1976:268) which, as is the case with banded linear decoration on Woodland pottery, is applied along and across a one-dimensional line axis. Patterns are then classified according to the type of motion by which these parts are repeated, resulting in the following seven classes: (1) translation, (2) bifold rotation, (3) horizontal reflection, (4) vertical reflection, (5) horizontal and vertical reflection, (6) slide reflection and (7) alternate rotation and vertical reflection (Washburn and Crowe, 1988:52–56, Appendix 2). Although traditionally couched within a structuralist ontology defined by cognitive templates (e.g., Hodder, 1982), an attempt was made to consider design symmetry within the Peircean semeiotic. Like other aspects of form and decoration in pottery, design symmetry was treated as a phenomenon intimately shaped and directed by the products and conditions of its materialisation. In other words, it is the possibilities afforded by and encapsulated within the materials themselves, along with the contexts of their creation and use that were seen to provide objects for interpretants and enable certain forms of relationality while excluding others. Malafouris (2005:58) adopts a similar perspective when he argues that while we are certainly able to evoke an idealised image of things in the world (e.g., through the use of language), material culture makes it possible for the mind to think through things, in action, without deferring to abstract mental representation. Artisanal practices were seen in this study to be influenced as much by the pervasiveness of extant designs (working as indexical legisigns) as they were by any preconceived mental template (or symbolic legisign) to which such designs must faithfully adhere.

Owing to limitations of space, only a brief summary and interpretation of the data is presented here. With regard to vessel morphology, distinctive patterns of lip construction (thickness and form) and overall rim profile were noted: relatively thick vessels with rounded lip surfaces tended to occur more frequently on Western Basin vessels while thinner vessels with flat lip surfaces were more common to Iroquoian collections. Attributes of vessel decoration, on the other

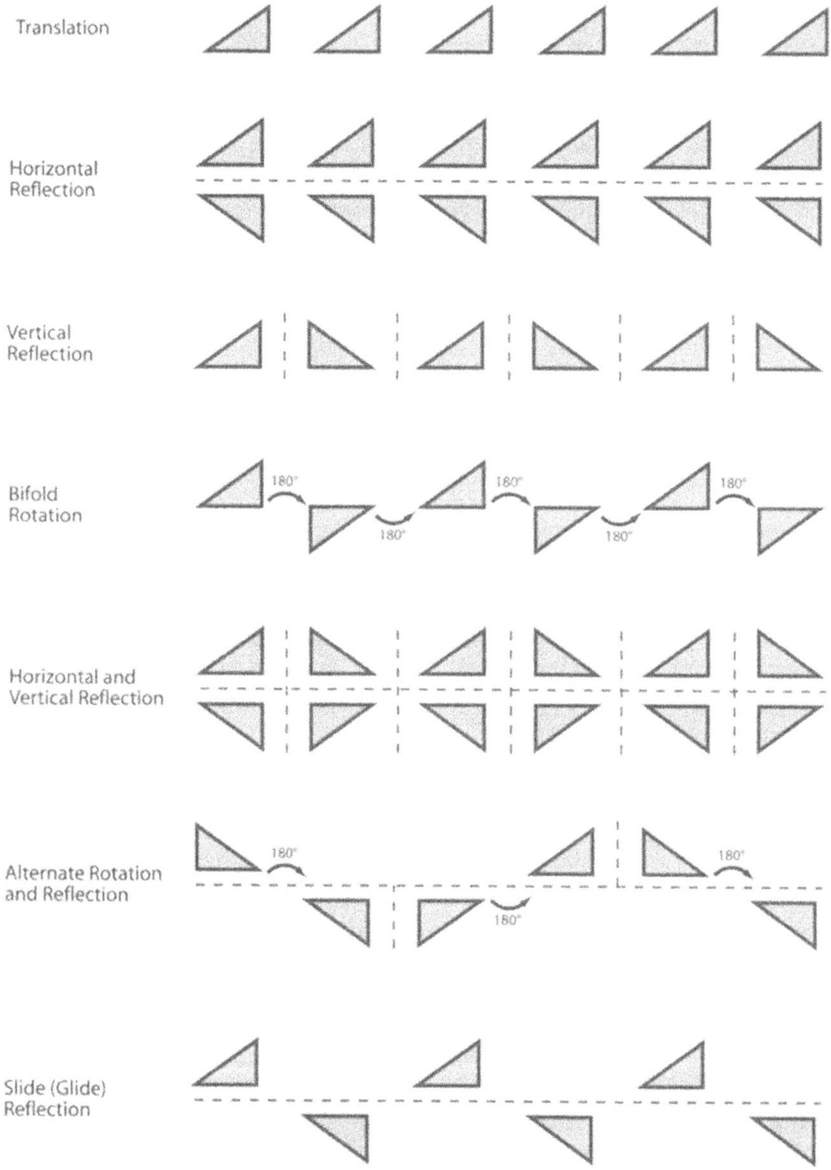

Fig. 10.3 The seven classes of one-dimensional symmetry

hand, reveal supra-regional trends only insofar as Iroquoian vessels are concerned, with Western Basin vessels instead being characterised by a broad array of attribute states, particularly on exterior bands. Having said this, there would appear to be a certain degree of diversity in decorative practices at both Iroquoian and Western Basin settlements. Given the above, we might suggest that where aspects of morphology are affected, there is a strong tendency for both Western

Basin and Iroquoian vessels to operate as dicent indexical legisigns. In this sign possibility, a general rule exists whereby each manifestation of vessel form, as a dicent (indexical) sinsign, more specifically a 'replica', clearly specifies the ground between object and sign. Here we might suggest that the semeiotic functioning of pottery form among these groups more forcefully directs the subsequent conventions governing its materialisation. Where decoration is concerned, however, the same 'significate effects' might be seen to operate among Iroquoian but not Western Basin potters. Indeed, Western Basin decorative practices were likely attuned to a semeiotic mediation involving rhematic indexical legisigns, where configurations of decorative bands, as rhematic indexical sinsigns (replicas), point to 'play' or possibility.

With regard to symmetry, both Western Basin and Iroquoian vessels displayed intra-band configurations consisting of simple repetition of elements (translation), while ratios of different symmetrical configurations between bands failed to illuminate any significant differences between Iroquoian and Western Basin practices. Of interest, however, was the preference among Iroquoian potters for symmetrical classes which involve "translation" or plain (asymmetrical) patterns on vessel necks, which differs substantially in extent from the spectrum of classes found on Western Basin neck surfaces. Indeed, all seven classes of one-dimensional infinite symmetry are found at the Western Basin Krieger site, while five of the seven classes are found in the Dymock collection. Like those properties associated with vessel decoration, the overall patterning which emerges from an analysis of intra-band symmetry on vessel necks suggests a form of semeiosis whereby indexical legisigns again cleave along rhematic (Western Basin) versus dicent (Iroquoian) lines.

Ultimately, my analyses suggested that Iroquoian potting practices during Late Woodland times were mediated by a relatively coordinated form of material agency. This pattern is at variance with the data from Western Basin collections which is illustrative of a greater degree of heterogeneity in vessel decoration and symmetry and suggests that a sense of choice was engendered among these groups. Elsewhere (Watts, 2007, 2008) I have suggested that these data point to key dissimilarities in the hybridised actor-networks and semeiotic articulations of Iroquoian and Western Basin 'pot/potters'. During the time period in question, Iroquoian groups would have been living in emergent, semi-sedentary villages with an economy increasingly oriented toward the production of maize. Accordingly, Iroquoian lifeways might have been suffused with a newfound sense of permanency, whereby daily practices were increasingly habitualised and enveloped in more intimate connections with local landscapes. It follows from such a view that the lives of things, owing to the more frequent appearance of both artisan and media within the same nodes of a network, would have worked toward a semeiosis based on causality rather than possibility. Pottery, like other objects stationed in particular spaces, might have been more firmly entrenched within Iroquoian actor-networks and, as such, may have been disclosed to potters through semeiotic functioning as a series of more factually-based interpretants.

For Western Basin groups, a more diversified economic lifeway coupled with the higher degree of mobility found in such an arrangement, suggests the corpus of pottery present at any one site would not have enjoyed the same conditions of materialisation or fixity within actor-network nodes as were likely found within Iroquoian settlements. Although still a cottage industry, the fleeting nature and lower population aggregates associated with individual sites suggests young potters, who learn by imitation, would not have been exposed to the same craft dynamics as their Iroquoian peers. With regard to duration, the absence of a fixed household unit may have impeded the development of a lineal craft tradition, particularly if pottery was not seen and practically experienced by users within long-term settings. Add to this an absence of strong potting communities and we can see how a semeiosis based on fluidity might lead to a pastiche of formal and decorative conventions and how the interpretants of such may have been imparted to artisans as mere possibilities.

Conclusions

This chapter has attempted to discuss non-anthropocentric forms of agency through an exploration of the Peircean semeiotic. I have argued, following Peirce's notions of synechism, semeiosis and sign possibilities that conceptions of material agency find a more sympathetic interpretive milieu if studied as part of a continuous process of sign generation and interpretation. The object-sign-interpretant triad helps us to develop an argument for a sovereign process of semeiotic functioning wherein meaning is neither resident in the mind nor contingent on human consciousness. Indeed, the very existence of humans and non-humans as meaningful entities inheres in the various connective properties that 'ground' object and sign. Thus, one of the more substantive implications of the Peircean semeiotic is that people and things, including their interests, are conjoined through the principle of semeiotic mediation. As relational entities, both depend on their being part of the sign process, on semeiosis and I have endeavoured above to point out how this process can be affected by objects of reality and their materialisation.

And here is perhaps one of the more important ramifications of the semeiotic for notions of material agency: signs do not simply transport information from one locus to another. They act as interlocutors, not vehicles and without the requisite mediate contact between object and representation, there is no triadic semeiosis. There are many negative consequences to a view of material culture which holds that things 'act' but only inasmuch as they provide a storehouse of symbols for an insular, subject-based conception of the world. If examined as part of a semeiotic frame, it becomes easier to distance ourselves from such a myopic outlook. Minimally, we can see materials functioning as 'lower-order' iconic or indexical signs outside the scope of cognitive representation. When brought into a triadic alignment, these same materials actively specify the sign's possibilities and are reflexively implicated by the interpretant. This is what enables representation to first take place—not the 'meaning content' of a thing but its material presence. Moreover, through its emphasis on continuity, the semeiotic allows us to overcome one of the great schisms native to modernity, that of the material/ideal divide. While Peirce recognised this distinction, to be sure, his semeiotic does not separate cognition from the effects of 'being in the world'. It simply highlights the triadic unity of experience and delimits the manners in which people and things interact in a universe perfused with signs.

Acknowledgments I would very much like to thank Carl Knappett and Lambros Malafouris for inviting me to contribute a chapter to this volume. My thinking on this topic has benefitted greatly from their advice and observations. I am also indebted to Robert Preucel, who very kindly read and commented upon an earlier draft of this chapter. Any faults or omissions in this work, however, are entirely my own. Finally, the research described herein was supported by a Social Sciences and Humanities Research Council of Canada Doctoral Fellowship, as well as several University of Toronto Open Fellowships.

References

Akrich, M., and Latour, B., 1992, A Summary of a Convenient Vocabulary for the Semiotics of Human and Nonhuman Assemblies. In *Shaping Technology/Building Society: Studies in Sociotechnical Change*, edited by W. Bijker and J. Law, pp. 259–264. MIT Press, Cambridge, Massachusetts.

Bauer, A.A., 2002, Is What You See All You Get? Recognizing Meaning in Archaeology. *Journal of Social Archaeology* 2:37–52.

Boast, R., 1997, A Small Company of Actors: A Critique of Style. *Journal of Material Culture* 2:173–198.

Capone, P.W., and Preucel, R.W., 2002, Ceramic Semiotics: Women, Pottery, and Social Meanings at Kotyiti Pueblo. In *Archaeologies of the Pueblo Revolt: Identity, Meaning, and Renewal in the Pueblo World*, edited by R.W. Preucel, pp. 99–113. University of New Mexico Press, Albuquerque.

Coben, L., 2006, Other Cuzcos: Replicated Theaters of Inka Power. In *Archaeology of Performance: Theaters of Power, Community and Politics*, edited by T. Inomata and L. Coben, pp. 223–259. AltaMira Press, Walnut Creek, California.

Costall, A., 1995, Socializing Affordances. *Theory and Psychology* 5:467–481.

Costall, A., 1997, The Meaning of Things. *Social Analysis* 41:76–85.

Dobres, M.-A., and Robb, J.E., 2000, Agency in Archaeology: Paradigm or Platitude. In *Agency in Archaeology*, edited by M.-A. Dobres and J.E. Robb, pp. 1–17. Routledge, London.

Eco, U., 1976, *A Theory of Semiotics*. Indiana University Press, Bloomington.

Gardin, J.-C., 1992, Semiotic Trends in Archaeology. In *Representations in Archaeology*, edited by J.-C. Gardin and C. Peebles, pp. 87–104. Indiana University Press, Bloomington.

Gell, A., 1998, *Art and Agency: An Anthropological Theory*. Clarendon Press, Oxford.

Gosden, C., 2005, What Do Objects Want? *Journal of Archaeological Method and Theory* 12:193–211.

Greimas, A.J., 1966, *Sémantique Structurale*. Librairie Larousse, Paris.

Hawkes, T., 1977, *Structuralism and Semiotics*. University of California Press, Berkeley.

Hodder, I., 1982, Sequences of Structural Change in the Dutch Neolithic. In *Symbolic and Structural Archaeology*, edited by I. Hodder, pp. 162–177. Cambridge University Press, Cambridge.

Hodder, I., 1989, This is Not an Article About Material Culture as Text. *Journal of Anthropological Archaeology* 8:250–269.

Hodder, I., 1992, Material Practice, Symbolism and Ideology. In *Theory and Practice in Archaeology*, edited by I. Hodder, pp. 201–212. Routledge, London.

Hookway, C., 1985, *Peirce*. Routledge and Kegan Paul, London.

Keane, W., 2003a, Self-Interpretation, Agency, and the Objects of Anthropology: Reflections on a Genealogy. *Comparative Studies in Society and History* 45:222–248.

Keane, W., 2003b, Semiotics and the Social Analysis of Material Things. *Language and Communication* 23:409–425.

Keane, W., 2005, Signs are not the Garb of Meaning: On the Social Analysis of Material in Things. In *Materiality*, edited by D. Miller, pp. 182–205. Duke University Press, Durham, North Carolina.

Knappett, C., 2002, Photographs, Skeumorphs and Marionettes: Some Thoughts on Mind, Agency and Object. *Journal of Material Culture* 7:97–117.

Knappett, C., 2005, *Thinking Through Material Culture: An Interdisciplinary Perspective*. University of Pennsylvania Press, Philadelphia.

Latour, B., 1993, *We Have Never Been Modern*. Translated by C. Porter. Harvard University Press, Cambridge, Massachusetts.

Law, J., 1999, After ANT: Complexity, Naming and Topology. In *Actor Network Theory and After*, edited by J. Law and J. Hassard, pp. 1–14. Blackwell Publishers, Oxford.

Lele, V.P., 2006, Material Habits, Identity, Semeiotic. *Journal of Social Archaeology* 6:48–70.

Lenoir, T., 1994, Was the Last Turn the Right Turn? The Semiotic Turn and A. J. Greimas. *Configurations* 2:119–136.

Malafouris, L., 2005, The Cognitive Basis of Material Engagement: Where Brain, Body and Culture Conflate. In *Rethinking Materiality: The Engagement of Mind with the Material World*, edited by E. DeMarrais, C. Gosden and C. Renfrew, pp. 53–62. McDonald Institute for Archaeological Research, Cambridge.

Olsen, B., 2003, Material Culture after Text: Re-Membering Things. *Norwegian Archaeological Review* 36:87–104.

Parmentier, R.J., 1994, *Signs in Society: Studies in Semiotic Anthropology*. Indiana University Press, Bloomington.

Parmentier, R.J., 1997, The Pragmatic Semiotics of Cultures. *Semiotica* 116:1–115.

Peirce, C.S., 1958–1965, *The Collected Papers of Charles Sanders Peirce*. 8 volumes. Edited by C. Hartshorne, P. Weiss, and A.W. Burks. Harvard University Press, Cambridge, Massachusetts.

Peirce, C.S., 1967, Manuscripts in the Houghton Library of Harvard University, as identified by Richard Robin. *Annotated Catalogue of the Papers of Charles S. Peirce*. University of Massachusetts Press, Amherst.

Peirce, C.S., 1977, *Semiotics and Significs*. Indiana University Press, Bloomington.

Preucel, R.W., 2006, *Archaeological Semiotics*. Blackwell Publishers, Oxford.

Preucel, R.W., and Bauer, A.A., 2001, Archaeological Pragmatics . *Norwegian Archaeological Review* 34:85–96.

Savan, D., 1989, *An Introduction to C. S. Peirce's Full System of Semeiotic*. Toronto Semiotic Circle, Monograph No. 1.

Schnapp, J., Shanks, M., and Tiews, M., 2004, Archaeology, Modernism, Modernity. *MODERNISM/modernity* 11:1–16.

Sebeok, T.A., and Uniker-Sebeok, J., editors, 1992, *Biosemiotics. The Semiotic Web 1991*. Mouton de Gruyter, New York.

Sebeok, T.A., Hoffmeyer, J., and Emmeche, C., editors, 1999, Biosemiotica. *Semiotica* 127.

Shanks, M., and Tilley, C., 1987, *Re-Constructing Archaeology: Theory and Practice*. Cambridge University Press, Cambridge.

Shepard, A.O., 1976, *Ceramics for the Archaeologist*. Publication Number 609, Carnegie Institution of Washington, Washington, DC.

Short, T.L., 2004, The Development of Peirce's Theory of Signs. In *The Cambridge Companion to Peirce*, edited by C. Misak, pp. 214–240. Cambridge University Press, Cambridge.

Thomas, J., 2004, *Archaeology and Modernity*. Routledge, London.

Thomas, J., 2007, The Trouble with Material Culture. *Journal of Iberian Archaeology* 9/10:11–23.

Washburn, D.K., and Crowe, D.W., 1988, *Symmetries of Culture: Theory and Practice of Plane Pattern Analysis*. University of Washington Press, Seattle.

Watts, C., 2007, From Purification to Mediation: Overcoming Artifactual 'Otherness' with and in Actor-Network Theory. *Journal of Iberian Archaeology* 9/10:39–54.

Watts, C., 2008, *Pot/Potter Entanglements and Networks of Agency in Late Woodland Period (ca. AD 900–1300)* Southwestern Ontario, Canada. British Archaeological Reports, International Series, Archaeopress, Oxford (forthcoming).

Whitridge, P., 2004, Whales, Harpoons, and Other Actors: Actor-Network Theory and Hunter-Gatherer Archaeology. In *Hunters and Gatherers in Theory and Archaeology*, edited by G. M. Crothers, pp. 445–474. Occasional Paper No. 31, Center for Archaeological Investigations, Southern Illinois University, Carbondale.

Chapter 11
When ANT meets SPIDER: Social theory for arthropods

Tim Ingold

Deep in the woods, amidst the undergrowth and detritus of a forest floor, two distinguished arthropods – renowned in the animal kingdom for their ingenuity and technical accomplishments – have struck up a conversation. One is ANT, the other is SPIDER. Both being philosophically inclined, their concern is to understand the world and their place in it. On this particular occasion, it is ANT's turn to open the debate.

'We ants', he declares, 'are not isolated individuals. Our brains may be no bigger than pin heads, yet we can achieve great things. Our nests are monumental mounds and our roads are highways through the forest, overrunning everything in their path. We can accomplish these feats because we collaborate. We live together in colonies, many thousand strong, sharing our food and work. In a word, we are the most *social* of insects'.

SPIDER, more solitary by nature, finds the idea of life in a colony hard to grasp. She admits that she would be more inclined to eat others of her kind than to work with them. Curious to know what it means to be social, she resolves to press ANT on the issue. 'In the course of your activities', she remarks, 'you have to deal with all sorts of things. I have seen you dragging worms and bugs that you have killed for food to your nests, along with building materials like twigs, pine needles and leaves, often many times your body size. I have seen you "touching up" aphids and licking the honeydew from their bodies. And I have seen you picking up and carrying around the larvae of your own kind. Tell me, do you have social relations with these things or only with mature members of the colony like yourself?'

'Now there, my dear SPIDER', replies ANT, 'you have touched on an issue that has been the source of some controversy in the formicoid world and I have to confess that my own views on the matter are somewhat unorthodox. To cut a long story short, there have up to now been two schools of thought. According to one school, we should think of the colony as a functioning totality that is more than the sum of its parts – a sort of super-organism – within which the life

T. Ingold
Department of Anthropology, School of Social Science, University of Aberdeen, Aberdeen, UK
e-mail: tim.ingold@abdn.ac.uk

C. Knappett, L. Malafouris (eds.), *Material Agency*,
DOI: 10.1007/978-0-387-74711-8_11, © Springer Science+Business Media, LLC 2008

of every individual is entirely given over to the greater good of the collectivity. According to the other school, what we call "the colony" does not correspond to any real, concrete entity. We merely use the term as shorthand for what, in reality, is a vast aggregation of individuals, each driven by those basic instincts with which it has been innately endowed. My own view, however, is that we should characterise the colony, in the first place, in terms not of its membership or composition but of what is actually going on there. Every colony is a hive of activity. And if we follow the lines of activity, we find that they can be traced back neither to a single, collective super-organism nor to a plurality of individual organisms. Rather, to trace the lines of activity is to describe a vast network, in which any individual appears as but a particular node. Every ant in the colony is part of the action and carries it forward in its own way; it is, if you will, an *act*-ant'.

'So if you want to assign responsibility for what is going on', interjects SPIDER, 'you could not lay it at the door of the individual or the collectivity. It is rather spread around the entire network'.

ANT waves his antennae in approval. 'Exactly so. That's why I say that the individual act-ant is not an agent. Rather, agency – that is, what makes things happen – is *distributed* throughout the network'.[1]

'That is all very well', retorts SPIDER, 'but you have still not answered my original question. You speak of the colony as a network of *act*-ants. But can the network also include *non*-ants? Can non-ants also have social lives?' '

Absolutely', ANT continues. '*Anything* can belong to the network, whether ant or non-ant. It is on precisely this point that I take issue with my colleagues. They seem to think there is something about being an ant – some essential anthood – that sets them apart from other creatures, in a separate world of *anture* as distinct from the material world of *nature* in which the existence of all other creatures is confined. Social relations, they claim, are not natural but *antural.* But the world I inhabit comprises both act-ants and non-ants, including such things as pine-needles, aphids and larvae. I insist that these things are not just passive objects. I am bound up in relations with them, as I am with my fellow ants. They, too, are part of the network. And they are caught up in it just as flies, my dear spider, are caught up in your web'.

'But there you are surely wrong', exclaims SPIDER. 'The lines of my web are not at all like those of your network. In your world there are just entities – bits and pieces of diverse kinds that are brought together or assembled so as to make things happen. Every "relation" in the network, then, is a connection *between* one entity and another. As such, the relation has no material presence. For the materiality of the world, in your view, is fully comprehended in the entities connected. The lines of

[1] As John Law and Annemarie Mol put it (Chapter 4, this volume), 'entities ... enact each other. In this way of thinking agency becomes ubiquitous, endlessly extended through webs of materialised relations'. What matters, then, is not what these entities are, but what they do, and reciprocally, what is done to them. Curiously, Law and Mol claim that the English language makes it difficult to express a condition intermediate between doing and being done to, despite the rich vocabulary of vernacular terms and phrases such as 'minding', 'keeping watch' and 'looking after'.

my web, to the contrary, are themselves spun from materials exuded from my own body and are laid down as I move about. You could even say that they are an extension of my very being as it trails into the environment – they comprise, if you will, my "wideware".[2] They are the lines *along* which I live and conduct my perception and action in the world.[3] For example, I know when a fly has landed in the web because I can feel the vibrations in the lines through my spindly legs and it is along these same lines that I run to retrieve it. But the lines of my web do not *connect* me to the fly. Rather, they are already threaded before the fly arrives and set up through their material presence the conditions of entrapment under which such a connection can potentially be established'.

SPIDER's account reminds ANT of an incident that took place during his winged mating flight, when he very nearly became caught in a spider's trap. It was touch and go but after a sticky experience he had eventually managed to break free. Was it the web, however, or the spider that had ensnared him? Wondering about this, ANT comes to the conclusion that 'it was, of course, both the spider and the web or what we might regard as a *hybrid* entity, the "spider-web", formed by their conjunction'. But there is more, as ANT goes on to explain. 'The web cannot function as a trap unless it is supported. In fact it was hung from lines attached to the twigs of bushes and to grass stems. So it was the way in which the spider, the web, the stems and the bushes all came together in the network, at that particular moment, that led to my nearly ending up as the spider's dinner'.

On hearing the word 'hybrid', SPIDER's legs begin to twitch nervously. She dislikes the term and has reservations about the way it has been promulgated by ANT and his confabulators.[4] 'Your talk of hybridity', she responds tetchily, 'entirely misses the point. You imagine a world of entities – spider, web, stems, twigs and so on – which are assembled to comprise the necessary and sufficient conditions for an event to happen. And you claim that the agency that "causes" this event is distributed throughout the constituents of the assemblage. My point, however, is that *the web is not an entity*. That is to say, it is not a closed-in, self-contained object that is set over against other objects with which it may

[2] The notion of 'wideware' is taken from Andy Clark (Chapter 1, this volume). 'The relation between the biological organism and the wideware', writes Clark, 'is as important and intimate as that of the spider and the web'. Elsewhere, art historian James Elkins draws on the metaphor of the web to describe the 'skein of vision' within which every human being catches the objects of his or her attention (or is alternatively caught). 'I am not the spider who weaves the web, and I am not even the fly caught in the web: I am the web itself, streaming, off in all directions with no center and no self that I can call my own (Elkins 1996:75).

[3] The same could be said of the trails or paths made by human walkers. Of such a trail, we might ask (as do Law and Mol, Chapter 4, this volume) 'where does this path come from, and where might it lead?' This is not an appropriate metaphor, however, for tracing the connections between entities in a network, as Law and Mol intend.

[4] An example can be found in Chapter 5 (this volume) in which Owain Jones and Paul Cloke speak of 'the ways in which non-human organisms and materials contribute to the networked agencies of hybrid collectifs'. Likewise Tom Yarrow, in Chapter 7, refers to 'the hybrid networks of people and things in which different kinds of "actants" are conjoined'.

then be juxtaposed or conjoined. It is rather a bundle or tissue of strands, tightly drawn together here but trailing loose ends there, which tangle with other strands from other bundles. For the twigs or stems to which I attach these trailing ends are themselves but the visible tips of complex underground root systems. Every plant, too, is a living tissue of lines. And so, indeed, am I. It is as though my body were formed through knotting together threads of life that run out through my many legs into the web and thence to the wider environment. The world, for me, is not an assemblage of heterogeneous bits and pieces but a tangle of threads and pathways. Let us call it a *meshwork*, so as to distinguish it from your *network*. My claim, then, is that action is not so much the result of an agency that is distributed around the network, but emerges from the interplay of forces that are conducted along the lines of the meshwork'.[5]

As ANT and SPIDER are conversing on the forest floor – surrounded by what ANT (the network builder) perceives as an assortment of heterogeneous objects and what SPIDER (the web weaver) perceives as a tissue of interlaced threads – something else is going on in the air above their heads. A couple of butterflies are dancing. 'Observe', says ANT, 'how in its fluttering, each butterfly responds to the movements of the other. We might even call it a "dance of agency".[6] Clearly, the butterflies are interacting in the air, just as we act-ants interact on the ground in the acrobatics of our collaboration'.

'But have you', asks SPIDER, 'given any thought to the air itself? The butterfly's flight is made possible, thanks to air currents and vortices partly set up by the movement of its wings. Similarly, the fish in the river is able to swim, sometimes at remarkable speed, because of the way it creates eddies and vortices in the water through the swishing of its tail and fins.[7] But what sense would it make to say that the air, in the first case, is a participant in the network, with which the butterflies dance as they do with one another; or in the second case, that the fish dances with water as it might with other fish in the shoal? Indeed it would make no sense at all. Air and water are not objects that act. They are material media in which living things are immersed, and are experienced by way of their currents, forces and pressure gradients. True, it is not the butterfly alone that flies but butterfly-in-air and not the fish alone that swims but fish-in-water. But that no more makes the butterfly a fly-air hybrid than it makes the fish a fish-water

---

[5] On the distinction between network and meshwork, see Ingold (2007: 80–2).

[6] The notion of the 'dance of agency' is taken from Lambros Malafouris (Chapter 2, this volume). In such a dance, Malafouris explains, there must be an equality or symmetry between the two partners. As the following argument makes clear, however, Malafouris is misguided in applying this notion to the relation between the potter and wet clay. For the potter and the clay are *not* equal partners. The clay is to the potter as air is to the butterfly, water to the fish, and the web to the spider. As such, it constitutes the ground for interaction, but is not an interactant.

[7] Andy Clark (Chapter 1, this volume) illustrates this point with the example of the tuna fish. 'The real swimming machine', he suggests, 'is thus the fish *in its proper context*: the fish plus the surrounding structures and vortices that it actively creates and then maximally exploits'. The 'proper context', in this case, is a fluid material medium with its pressure gradients and lines of force. It is not an assemblage of discrete material objects.

hybrid. It is simply to recognise that for things to interact they must be immersed in a kind of force-field set up by the currents of the media that surround them. Cut out from these currents – that is, reduced to objects – they would be *dead.* Having deadened the meshwork by cutting its lines of force, thus breaking it into a thousand pieces, you cannot pretend to bring it back to life by sprinkling a magical dust of "agency" around the fragments. If it is to live, then the butterfly must be returned to the air and the fish to the water'.

'And I', SPIDER goes on, 'must return to my web. For I have to say that what air is for the butterfly and water is for the fish, my web is for me. I cannot fly or swim, but I can weave a web and exploit its properties of stickiness, tensile strength and so on to run around and catch flies. I may dance the tarantella with the fly that alights on my web, but the web itself is not a dancing partner. It is not an object that I interact with, but the ground upon which the possibility of interaction is based. The web, in short, is the very condition of my agency. But it is not, in itself, an agent'.

'That, if I may say so', interjects ANT, 'is a very arachno-centric viewpoint. Presumably, by your same argument, if you were a fly you could also claim to be an agent, and if you were an ant like me, you could claim to be an agent too. How many legs, I wonder, do you need to qualify as an agent: six, eight, a hundred? Our mutual acquaintance the centipede would indeed do very well. With so many legs he must be a truly powerful agent'.

'You jest of course, my dear ANT', responds SPIDER. 'Nevertheless to your question I would answer: at least four! For although I would be prepared to admit to the agency of our four-footed friends, the rat and the mouse, I would draw the line at bipedal humans. You may be an agent from your formicoid perspective and I from my arachnid one, but from the perspective by which humans distinguish themselves from all other creatures, it is impossible to see how they could exercise any agency at all. On one occasion, I dangled inconspicuously from the ceiling of one of their so-called "classrooms" and overheard a human philosopher lecturing to others of his number. "I am a human subject", the man intoned. "*I know, therefore I am.* I know, and am, because I have a mind. That is what makes me human. And it is this, too, that enables me to act. Of course I have a body too, like every other creature. The spider has a body; so does the ant. But the spider and the ant are all body; there is no more to them than that. Though we may observe their behaviour, they cannot act. But *I* am not my body. I am a body *plus*. It is by the measure that I am *more* than my body that my humanity – along with the scope of my action – is defined".

'"Well", thought I silently to myself, as I swung from the end of my thread, "if that's where you imagine the essence of your humanity lies, then it is certainly not to be found in what you humans do. What you have been talking about is intelligence, a cognitive capacity to work things out in advance, in the head, prior to their implementation in the world. But *intelligence* is one thing; *agency* quite another.[8] It is a serious mistake to confuse the two". And I

[8] 'Intelligence', as Richard Harper, Alex Taylor and Micheal Molloy state (somewhat controversially) in Chapter 6, 'is a term only applicable to human beings'. But 'agency' is not.

remembered the story of the apocryphal centipede who, when asked how he managed to co-ordinate the movements of his hundred legs, found himself paralysed and starved to death. So long as he had acted unthinkingly, leaving his legs to look after themselves, there had been no problem. But as soon as he stopped to think intelligently about what he did, he could no longer act. His agency was thwarted'. More generally, a creature that could do nothing that had not been fully thought out in advance could never, in practice, do anything at all.

'We all know about the arrogance and stupidity of humans', laughs ANT in response, 'especially the philosophers among them who have nothing else to do in life than to think. If we could only reduce them in scale and put them to work in one of our nests, they would learn a thing or two! They would soon discover, as I have explained already, that agency is not exclusive either to ants or to non-ants but is distributed throughout the network formed by their collaboration. We need, in short, to establish a principle of symmetry, by which neither side of the ant/non-ant dichotomy is privileged over the other'.

'I do not want to accord a special privilege to ants or to spiders', responds SPIDER, 'let alone to human beings. Yet I cannot accept your principle of symmetry. The problem lies in your blanket category of the 'non-ant' which includes everything from grains of sand and dead leaf-matter to aphids and butterflies – and even humans! Our concept of agency must make allowance for the real complexity of living organisms as opposed to inert matter. It is simply absurd to place a grain of sand and an aphid on the scales of a balance and to claim that they are equivalent. They may weigh the same amount, but in terms of complexity they are poles apart. The key difference is that the aphid, animal that it is, has a nervous system – just as do you and I. When I crouch at the centre of my web, I am all a-quiver, just like the leaf of a tree in the summer breeze. I am sensitive to the slightest movement or vibration. What makes the difference between me and the leaf, however, is that every movement I make is also a movement of my *attention*. It is the attentiveness of this movement that qualifies it as an instance of *action* and, by the same token, qualifies me as an *agent*. To put it another way, the essence of action lies not in aforethought (as our human philosopher would claim) but in the close coupling of bodily movement and perception. But that is also to say that all action is, to varying degree, *skilled*. The skilled practitioner is one who can continually attune his or her movements to perturbations in the perceived environment without ever interrupting the flow of action.[9] But such skill does not come readymade. Rather, it *develops*, as part and parcel of the organism's own growth and

[9] As John Sutton explains (Chapter 3, this volume), 'the minutely adaptable exercise of embodied skills precisely requires an openness to and awareness of the specifics of a situation'. But this does not mean, as Sutton seems to think, that the 'grooved embodied engagement' of the practitioner is modulated by cognitive intervention or 'doing' by 'knowing'. Rather, practice that is skilled does not follow a groove at all but rather cuts its own groove as it is guided by perception.

development in an environment. Since agency calls for skill, and since skill arises through development, it follows that the process of development is a *sine qua non* for the exercise of agency. To attribute agency to objects that do not grow or develop that consequently embody no skill and whose movement is not therefore coupled to their perception, is ludicrous'.

Listening to this, ANT remains unimpressed. 'Well, you would say that, wouldn't you', he remarked caustically. 'You are SPIDER, and you stand for the proposition that *S*killed *P*ractice *I*nvolves *D*evelopmentally *E*mbodied *R*esponsiveness. I appreciate your views; they are indeed worth their weight IN GOLD (which is very little, I might add, since you are such a lightweight creature). But I am ANT. I stand for *A*ctor *N*etwork *T*heory. Not for nothing am I known as THE TOWER among arthropods. For my philosophy towers over yours'.

'You are indeed a master of lofty thoughts', admits SPIDER wearily. 'But I cannot, for the most part, understand a word of what you say'. And with that, she scuttles off.

References

Elkins, J., 1996. *The Object Stares Back: On the Nature of seeing*. Simon and Schaster, New York.

Ingold, T. 2007. *Lines: A Brief History*, London: Routledge.

Chapter 12
Agency, Networks, Past and Future

Sander E. van der Leeuw

Abstract In this contribution, I will first draw upon the other chapters of this book to summarise what are in my opinion the essential elements of their discussion on the concept of agency and on Actor Network Theory. Then I will argue the need to come to understand the process of invention and innovation and present some ideas about why this topic has not generally been given the importance that, in my eyes, it merits. Finally, I will try and develop a perspective that might indeed help us understand the process of intention and innovation, based on some of the ideas about agency presented in this book.

Introduction

While I was working on my PhD thesis "*Studies in the Technology of Ancient Pottery*" (1976), I had the good fortune to do much of it in dialogue with an experienced potter, Jan Kalsbeek, who had himself used the foot-driven wheel to manufacture a wide range of ceramic artefacts in a semi-industrial context before being appointed to the University of Leyden as a specialist in ancient ceramic technology.

At the time (1969–1976), the approach to ceramic analysis that we were working on was quite novel and demanded the development of new ways of thinking about ancient technology, about pottery manufacture and about ceramic analysis in archaeology. Our discussions were wide-ranging and often heated as they confronted two very different perspectives.

Jan Kalsbeek's perspective was that of the practitioner who tried to 'get into the mind of the ancient potters', to find out what they did to create the pots that we studied the broken remains of and why they did it. For that purpose, he used his own inimitable mixture of acute observational and analytical skills, knowledge about the potter's craft, extensive experiments in his workshop and

S.E.van der Leeuw
School of Human Evolution and Social Change, Arizona State University, Tempe, Arizona, US
e-mail: vanderle@asu.edu

C. Knappett, L. Malafouris (eds.), *Material Agency*,
DOI: 10.1007/978-0-387-74711-8_12, © Springer Science+Business Media, LLC 2008

descriptions of ceramic manufacture from the ethnographic and historical literature. I, on the other hand, was charged with presenting the results in such a way that they were understandable and acceptable to a scientific audience.

I realised quite a few years later that the heated nature of our discussions was in part due to the fact that we did not understand the inherent difference between our two perspectives. Whereas I, a freshly trained (positivist) scientist at the time, tried to write a coherent discourse about the how and why of the procedures used by ancient potters to shape and fire their wares in terms of chains of cause-and-effect and conscious choices, Jan verbalised the process of pottery making not in terms of cause-and-effect but in terms of possibilities and probabilities. Choices, in his view, were not necessarily conscious but often inherent in 'the way things were done' – in culturally determined 'know-how' anchored in they ways potters had learned their craft.

In other words, while Jan Kalsbeek looked at pottery making as a creative activity, in which the potter was faced with certain opportunities and challenges and used his skills to 'manage' both, I looked at the result of the potter's work and tried to explain how it came into existence. His perspective was directed towards understanding what enabled the future existence of the pots, whereas mine was focused on understanding the origins of the pottery I had in front of me. The potter in Jan phrased his conclusions in 'a priori' terms – positioning himself at a time before the pottery was created and looking forward towards its creation. I worded my conclusions in 'a posteriori' terms, positioning myself as the observer of the – already made and broken – pots and trying to explain how they were made and broken. Jan was thinking in terms of *prediction of what might have happened, what did happen and what did not happen* whereas I was working towards an *explanation of what did happen*. His language was that of uncertainty, probability and scenarios of possible future events, whereas mine was that of certainties, cause-and-effect and models of past events. Needless to say that his objection to my writings was, time and again, that I oversimplified and distorted the essence of what he was trying to express.

Human History and 'Natural History'

The implications of this difference in perspective are an important source of confusion in much interdisciplinary research that combines, as we often do in our discipline, approaches from the sciences on the one hand, and perspectives from the humanities on the other. Whereas the life sciences, in what used to be called 'natural history', tend to construct their vision of the past by referring to their understanding of the present, the historical sciences, dealing with human events, tend to explain their vision of the present by referring to the past. Among other things, this is reflected in the respective disciplines' perspective

Table 12.1 Contrasting 'human history' and 'natural history'

Human history ('a priori') approach	Natural history ('a posteriori') approach
Main interest in past	Main interest in present
The present understood in terms of the past	The past understood in terms of the present
Time and process irreversible	Time and process reversible, cyclical or reproducible
Accentuates differences	Accentuates similarities
Case studies	Generalisations
Language of probabilities	Language of certainties
Open-ended categories	Closed categories
Contingency	Causality
Focus on inter-scale interaction	Focus on intra-scale interaction
Scenarios	Models

on time. Whereas historians studying the actions of people have never really wavered in their vision of time as an irreversible 'arrow', natural and life scientists have for a long time approached the study of their domain by assuming the absence, the reversibility or the cyclicality of the temporal dimension in which processes occurred. This has only changed recently, when the complex adaptive (self-organising) systems paradigm was introduced. Table 12.1 summarises some of the implications of this difference in approach.

Archaeology is indeed profoundly affected by this opposition. Whereas the more distant pre-historic past is studied according to the canons of the natural and life sciences, the last five thousand years (the historic past) are studied with approaches coming from the humanities. The period in between (proto-history) is from this perspective a kind of confused 'no man's land', in which both approaches are being used. But, of course, one also finds 'historical' interpretations of pre-history and 'scientific' interpretations of more recent archaeology. Indeed, very often, authors are unaware of this aspect of their perspective, especially when socio-environmental dynamics are discussed.

The reader will have understood that this opposition has, in my opinion, also important ramifications for the discussions in this book, and in particular where they touch on the interaction between human and non-human agency, on action and re-action, and therefore on the debate around agency that is the subject of the preceding chapters.

In reading these chapters as a relative outsider to the domain, I was inspired by the many possibilities offered by the convergence of approaches presented in this volume to anyone interested in improving our understanding of the role of material culture in society, of human cognition but also of the dynamics of invention and innovation. In the next few pages, I will summarise some of the new strategies I see emerging.

What is New?

As mentioned, I am a newcomer to the discussions presented in these pages and when I ask 'what is new?', that is not meant as a starting point for a detailed discussion of what exactly in this inspiring volume has been said before and what has not. I think one of the most interesting aspects of the book is that it mingles and combines a number of different domains and approaches that have a history of their own but do not have a history in common. This section is merely intended to summarise what in my eyes are some of the main characteristics and implications of the shift in perspective proposed by the authors of the foregoing chapters.

At the core of the book is a perspective on the relationship between people and material culture that accords a more active role to the latter. The initial move towards this perspective goes back some twenty years or so but "has been more a case of acknowledging the active [...] use of material culture by humans, rather than ascribing much dynamism to the artifacts themselves" (Knappett & Malafouris, this volume). The various authors in this volume go a step further, and open the 'Pandora's box' of *assigning an active role to the artefacts themselves*, in particular in the context of human *cognition* (Clark, this volume, Chapter 1). Artefacts thus become 'agents' in the cognitive process just like human beings. This also changes the presumed role of the human brain, fuzzing the distinction between perception, cognition and action and emphasising action as the core focus of the brain. In Clark's words (infra, p.10):

> Perception is itself tangled up with possibilities for action and is continuously influenced by cognitive contextual and motor factors. It need not yield a rich, detailed and action-neutral inner model awaiting the services of 'central cognition' so as to deduce appropriate actions. In fact, these old distinctions (between perception, cognition and action) may sometimes obscure, rather than illuminate, the true flow of effect. In a certain sense, the brain is revealed not as (primarily) the engine of reason or quiet deliberation, but as the organ of *environmentally-situated control*. Action, not truth and deductive inference, are the key organizing concepts.

However important it may be, the cognitive process is only one of the loci of interaction between humans and artefacts. Another is that of *material engagement*, the interactive process by which human beings create or transform artefacts. Malafouris (this volume, Chapter 2) makes this point and in the process highlights another very important aspect of the overall shift in perspective: agency and intentionality are *distributed, emergent* and *interactive* phenomena in which materials, tools and brains each have their own roles, locating the neural contribution as just one (important) element in a complex causal web spanning brains, bodies and world, a perspective that is closely (and not accidentally) linked to the 'a priori perspective' of active creators such as Jan Kalsbeek. The role of the brain thus resembles that of a mediator who attempts to negotiate between the opportunities and challenges of aims (e.g., the characteristics of the end product), materials and tools to achieve the end product. In that process, the interaction between *prior intention* (what one sets out to do) and *intention in action* (what one does in the material world) is crucial (Searle, 1983), but not always clear: "[a]ll intentional actions have intentions in action, but not all intentional actions have prior intentions", and "[i]n both cases, the intention – as an internal representational state – causes the agent's movement – as an external physical state in the world. The difference is that in the case of "intention in action" the internal intentional state and the external movement become indistinguishable". According to Malafouris, such "*intention in action*" is not an internal property [of human beings] but a component of extended cognition which is part of the action rather than preceding it. I will return to this point a little further down when making the distinction between *knowledge* and *know-how*.

In Chapter 3, Sutton discusses Malafouris' (2004) perspective on Material Agency in the cognitive context and positions himself somewhat differently concerning the relation between people and artefacts in the acquisition of knowledge and know-how and in the transition between them. But he also adds an important concept, that of '*cognitive biography*'. Following Appadurai (1986) and Kopytoff (1986), his core assertion is that "the outcome of any action is also determined by the diachronic trajectories of things [and, I would add, ideas and people] through time and space". This seems to imply that there are certain continuities in conception and realisation of artefacts through time and that the distributed actor-artefact network has what amounts to a 'memory'. But for our purposes, it may even be more important that it therefore does not seem possible to invent literally 'anything, at any time'. Existing material culture (and the concepts and relations it represents and instantiates) seems to constrain the range of inventions and innovations that may emerge.

The concept 'artefact' that is traditionally used in such discussions also needs to be *extended, to all observable non-human entities*, including sheep and trees (Law and Mol, this volume, Chapter 4; Jones and Cloke, this volume, Chapter 5), but in effect extended to all of the atmosphere, hydrosphere, geosphere and biosphere. In certain instances, in my opinion, it may also be extended to human beings, that is, wherever humans are the object of other human beings' actions. Artefacts, moreover, play many different roles, related to the different perspectives from which active humans are considering them. Law and Mol point out that in the 2001 foot-and-mouth disease crisis in Cumbria, for example, sheep played roles in a veterinary, an epidemiological, an economic and a farming perspective. *Artefacts have more than one significance in the distributed network of agents*, depending on their partners in interaction and in particular on the functions they fulfill (the roles they play) in the network. Each significance comes with its own markers, its own characteristics and its own place in the human actors' worldview, etc.

Chapters 4, 5 and 9 also introduce an extant approach in social science – Actor Network Theory (ANT) – that aims to come to grips with the multiple interactions between different agents playing out in any cognitive or material engagement process. Introduced by Callon (1986), Latour (1993) and others, this approach has championed non-human agency whilst at the same time rejecting the non-human/human distinction. The agency of non-humans is seen as an essential element in the relationship between the natural and the social and *nature is framed as both a real material actor and a socially constructed object*. This leads one to assume and describe partnerships between human and non-human actors in mutual constructions of an artefactual nature (Latour's 'hybrids'), in which agency is viewed as located among different actors (human or not), decoupled from subject-object distinctions and representing the collective capacity for action by humans and non-humans. As a consequence, particular attention is paid to: (1) the *investigation of the nature and dynamics of the hybrid network itself* and (2) the *semiotic connections between the diverse elements in socio-technical ensembles* rather than to the entities that interact.

In such a network, agency is not located in any one place and it is multi-temporal (the result is determined by the interaction of many temporal rhythms occurring simultaneously in the network). The temporal and the spatial dynamics are, of course closely related. "*The [...] comings and goings in the production of place can be seen as processes of [temporal] patterning*" (Jones and Cloke, this volume, Chapter 5). Or, in the words of Harrison, Pyle and Thrift (2004): "All kinds of things can come together in the world and, in that process of encounter and settling down into at least a short-term equilibrium, they can creatively produce new kinds of organizations that are greater than the sum of their parts". "[...] all places are not merely processes and narratives but whole ecologies of interrelating trajectories that settle into temporary local forms but which also have threads that weave through the local to the global in scale" (cited by Jones and Cloke, this volume, Chapter 5).

Of the last two chapters of this volume that I wish to briefly include in the discussion, Yarrow (this volume, Chapter 8) focuses on aspects of *the semiotic relationship, in cognition and engagement between the material world and the world of ideas*. He shows how the administrative 'capture' of material archaeological data impoverishes the information inherent in those data by reducing the material world to its symbolic and meaningful content in the process of 'fitting' them in formal categories. Conversely, different meanings may be derived from what is manifestly the same text as it enters different contexts. To summarise this in my own terms (to which I will return): *moving from the material world to the world of ideas reduces the number of dimensions in the hyperspace associated with each artefact, whereas implementing ideas increases the number of dimensions in the hyperspace*.

Knappett's (this volume, Chapter 9) starting point is the distinction between 'things' and 'objects'. Whereas the former exist in the material world, the latter exist in the world of ideas. Things, according to Brown (2003) are ambiguous, undefined and have a metaphysical presence, whereas objects are named, understood and transparent, materialising out of the amorphousness of things. Gosden (2004) adds that objects are alienable, quantifiable and dis-embedded from social relations, whereas things are inalienable, possess unquantifiable qualities and are embedded in social relations as part of artefact communities from which they are difficult to extract without making them lose much of their meaning. The question is: how do 'things' become 'objects' or vice versa?

Found, nondescript 'things' may be transformed into artworks and then become objects. Naming something with a word can make an object out of a material thing and the same is true of imaging: the image of a thing can transform it into an object. In other words, *things in the material world are carriers of potential information and it is their link with concepts in the world of ideas (knowledge) that transforms them into objects that do carry significant information.* Evidently, the relationship is complex but one aspect of it is that 'thingness' is beyond representation, poly-interpretable and therefore nullified by naming or imagining – by creating a direct relationship with the world of ideas. Conversely, once something is brought out of 'thing-ness', it requires an interpretive framework (a world view), sometimes backed up by a scaffolding structure, to keep it suspended in 'object-hood'.

Depending on the relationship which is studied, *intermediates* (such as images, pictographs, ideographs, spoken or written words), can be both objects and things – they *are objects relative to the artefacts they represent, and they are things with respect to the imagery, writing system or language that is their context.* However, the nature of the representation is different. Images and pictographs are direct representations of things (or indirect representations of concepts), whereas spoken or written words are indirect representations of things (or direct representations of concepts). In the evolution of writing, the evolution from indirect to direct representation of concepts can be seen as a gradual dispelling of the 'shadow' of the objects.

I would like to conclude this section by emphasising that it would be too easy to conclude from the distributed, networked nature of action and agency that artefacts have their own kind of intelligence. Harper, Taylor and Molloy (this volume, Chapter 7) argue (rightly, in my opinion) that that is a step too far. In their terms, even modern information *technology is not offering intelligence, it is only offering people [. . .] resources to act and think.* While that may be simplifying things (as it does not take the distribution of cognition and action in networks of human and non-human actors into account), it does point to a very real danger of the cognitive and communication revolution and provides the necessary counter-arguments.

The Need for a New Kind of Innovation Studies

We live in a world that is driven by invention and innovation but that has not always been so. In the 17th century, innovation was a 'dirty word'. The world order was deemed to be immutable; people behaved as their ancestors had done (or at least they believed they did and they often strived hard to meet that ideal) (Girard 1990).

Little by little, though, over the last three centuries, 'history' and 'tradition' ceded place to 'nature' as the concept invoked to explain the world order. We still speak in many instances of 'it is natural' when we wish to express the fact that we think that a kind of behaviour is in harmony with the world order. In the process, in the first part of the 19th century, history has become a discipline rather than the omnipresent way to explain what happened or what happens.

Simultaneously, we observe a growing *emphasis on the new rather than the old* – particularly during, and as a result of, the Enlightenment and the Industrial Revolution (Girard 1990). As science and technology gained in importance, the conceptual and instrumental toolkit of the (western) world grew exponentially, and in doing so enabled humanity to identify and tackle more and more challenges. But solutions always leave unexpected challenges in their wake and they require more problem-solving. Hence, since the Neolithic but particularly during the last three or four centuries, the following feedback loop has been accelerating (van der Leeuw 2007):

Problem-solving structures knowledge —> more knowledge increases the information processing capacity —> that in turn allows the cognition of new problems —> creates new knowledge —> knowledge creation involves more and more people in processing information —> that increases the size of the group involved and its degree of aggregation —> creates more problems —> increases the need for problem-solving —> problem-solving structures more knowledge . . . etc.

As a result, the number of inventions and innovations around us is increasing almost exponentially. This is clearly visible if one looks at the number of inventions that are patented in the industrial countries (cf. Fig. 12.1).

After the industrial and nuclear revolutions, we are now witnessing the silicon, information technology and communications revolutions, and the nanotechnology, biotechnology and cognitive revolutions are on the horizon, each of which is opening another whole new domain of knowledge, know-how and innovation. For the moment at least, there does not seem an end in sight to this acceleration of change in our world.

Can the world sustain this? The social divide, the information divide, the health divide and so many other divides are growing, mainly because a smaller and smaller portion of the world population controls these innovations and the fruits thereof. Although many inventions are at least potentially beneficial, their sheer number increases the chance of unhappy surprises (misuse, but also accidents and unexpected long-term consequences of innovations) at a similarly exponential rate.

In that context, it is in my opinion surprising that the scientific community has so little understanding of the process of invention and innovation itself. Generally, the world reacts a posteriori to innovations once they have been introduced. Could we not attempt to shift our stance from a re-active to a pro-active one and come to understand and guide the process of invention and innovation itself? That would put us in control rather than dealing with things after they have got out of hand and it would potentially allow us to accelerate the innovative process in those domains in which that is most needed and maybe slow it in others . . .

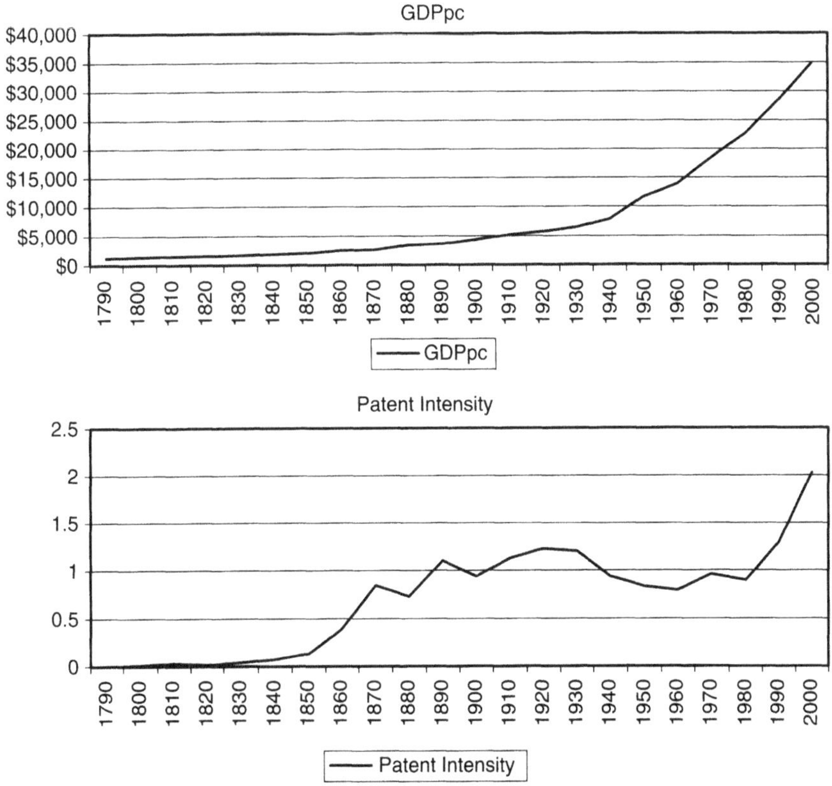

Fig. 12.1 Since around 1840, the number of patents accorded per capita in the US has increased substantively, the last two decades exponentially. The figure shows this increase, as well as the concomitant increase in GDP (reproduced with permission from the author, J. Lobo)

I have therefore been asking myself what has thus far held back our understanding of the process of invention and innovation, and that has been one of the main drivers for the ISCOM project that Carl Knappett refers to in his contribution. My tentative working hypothesis for the moment is that this lack of understanding is directly related to the fact that the majority of the scientific community has looked at invention and innovation using a positivist, scientific perspective such as the one I applied to ceramic studies. In essence, invention and innovation have mainly been studied ‘a posteriori’. In the introduction, I tried to make clear that from such a perspective, creation cannot be described or understood. Hence, we have left ‘invention’ completely to one side in innovation studies, relegating it to the domain of ‘personal creativity’ and we have focused uniquely on innovation, that is, on the ways in which an invention is adopted and spreads throughout a population.

What I would like to do in the remainder of this chapter is to apply the perspective of the creator – as represented in the introduction by Jan Kalsbeek – to the process of invention. The reason for this is that I think that this book has unwittingly lifted the tip of the veil by introducing a relational perspective that includes people, ideas, artefacts and resources in an interactive network.

From 'Being' to 'Becoming' and from Entities to Relations

At its core, this book proposes the (partial?) substitution of a relational perspective for the current emphasis on entities. In itself, that is not a new idea. In "*Tlön, Uqbar, Orbis Tertius*", for example, Jorge Luis Borges (1981) outlines what the consequences are of a perspective that excludes entities. One of the imagined languages of Tlön lacks nouns. Its central units are "impersonal verbs qualified by monosyllabic suffixes or prefixes which have the force of adverbs". In a world where there are no nouns — or where nouns are composites of other parts of speech, created and discarded according to a whim — and no *things*, most of Western philosophy becomes impossible. Without nouns about which to state propositions, there can be no a priori deductive reasoning from first principles. Without history, there can be no teleology. If there can be no such thing as observing the same object at different times [because change is omnipresent and permanent], there is no possibility of a posteriori inductive reasoning (generalising from experience). Ontology — the philosophy of what it means to *be* — is an alien concept.

The implications of such a shift in perspective, from a static to a dynamic world view, are taken up at considerable length in Ilya Prigogine's "*From Being to Becoming*" (1980). In such a perspective, we do not study 'what is' but 'what happens', we do not study 'entities' but 'processes', where change is assumed and stability needs to be explained, rather than vice versa. Some of its implications for archaeology have been described by McGlade and van der Leeuw (*Archaeology and nonlinear dynamics: new approaches to long-term change*, 1997). I ask the reader's indulgence for a brief summary of the most relevant ones.

One could compare the archaeologist to a prophet standing with his back to the future and looking into the past predicting the origin of known phenomena from a perspective diametrically opposite to that within which humans in the past made decisions about their future. That, of course, does not allow the archaeologist to understand the process of creation. To do so, he should, as it were, *travel back* in time and *look forward* with those whom s/he studies.

The success of the non-linear dynamic systems approach in representing hitherto inexplicable behaviour in a wide range of systems sheds a different light on one of the fundamental assumptions underlying most scientific explanation, the concept of causality. Unexpected system behaviour, which does not fit the usual idea of cause-and-effect, can 'suddenly' occur in an otherwise regular process. Moreover, from this perspective, similar causes can have widely divergent effects, but very different causes can, under certain circumstances, also have convergent effects.

The theoretical and methodological tools provided by this approach allow us to re–conceptualise the evolutionary behaviour of complex human systems from a non-linear dynamic perspective – one in which discontinuity and bifurcation are seen as intrinsic properties of open, dissipative systems. Such systems depend for their resilience and continued existence as much on innovative behaviour as they do on replication.

The focus on system resilience that is currently emerging in ecology, archaeology and other disciplines, attaches great importance to the capacity of a system to survive by changing 'in tune' with its environment. In doing so, it not only draws attention to the importance of innovative behaviour, but also to the fact that from this perspective, it is change which is assumed and stability which is questioned, rather than vice versa. As change is assumed to occur non-linearly in many cases, it is often important to have an idea of the change of change before change itself can be qualified and/or quantified at any point of the system's trajectory. Although very long-term models of the change of change often are virtually impossible to formalise in a testable manner, they ought to be present in the back of our mind when we formulate ideas concerning shorter-term developments. It probably serves an important purpose that such models are made explicit and subjected to scrutiny.

The focus on non-linear dynamics has also changed our perspective on space and time. Our awareness of non-linear phenomena puts density-dependent spatial phenomena in a new light and our conception of time is also changed, replacing a universal, external, analytical time dimension (radiometric dating in calendar years) by a highly individual non-linear phenomenon which is dependent on experiential and highly contextual attributes. The range of temporal rhythms occurring in any dynamic system are to a considerable degree, if not entirely, responsible for unexpected occurrences in the human/environmental co-evolution.

The Dynamics of Invention

Invention, and all creations, is a paradox. *Creation is at once the substantiation of form and the information of substance.* It occurs simultaneously and indivisibly in mind and in matter. On the one hand it is the creation of new ideas, which is instantaneous and implies change and uniqueness. But on the other, it is the introduction of a new process, a new form that fixes the transience of creation and thus implies control over matter and energy, and stability. The creation is subsequently imitated, making use of the dimensions 'discovered'. These two aspects of innovation are antithetical and define the relationship between 'innovation' and 'tradition' as a duality of perception.

Argueably, the set of solutions perceived *necessary* to the creation of a material object is structurally represented in memory as the set of relationships between a number of dimensions. But because that set is a post-facto construct, it is never *sufficient* to create another artefact, even one which is as closely similar as is theoretically conceivable. In the time elapsed since the first artefact materialised, both the observed context and its perception have inevitably changed, so that some more creation is needed: material culture is not replicated, it is re-created.[1] *In this sense, invention only exists as a 'special case' of change, from an a-posteriori perspective such as the archaeologist's.* From an a-priori perspective such as the artisan's, the distinction between innovation and repetition is only a quantitative one: every time, a jump needs to be made. *All* decision-making is invention as well as recognition.

In the last part of this chapter, I will try to show how the perspective on agency presented in this book might contribute to our understanding of the dynamics of invention. My point of departure is the basic distinction that Knappett (this volume, Chapter 8) makes between 'things' and 'objects':

> Evidently, the relationship is complex, but one aspect of it is that 'thing-ness' is beyond representation, poly-interpretable and therefore nullified by naming or imagining – by creating a direct relationship with the world of ideas. Conversely, once something is brought out of 'thing-ness', it requires an interpretive framework (a world view), sometimes backed up by a scaffolding structure, to keep it suspended in 'object-hood'.

In other words,

> [...] things in the material world are carriers of potential information, and it is their link with concepts in the world of ideas (knowledge) that transforms them into objects that do carry significant information.

The so-called 'scaffolding structure' (here used in the 'narrow' sense),[2] which enables the retention of the 'object-ness' of 'things', is always a distributed one, involving a network of associative relationships between different elements, both 'objects' and 'things' (including people and other living beings, other artefacts, tools, resources, etc.). That network can either reside entirely in the realm of ideas (and thus in the human mind), or it can have material components that 'activate' the human memory when the individual encounters them. These material components may be 'things' to which humans have 'delegated' certain marker functions (trees, rocks and the like, as in the case of animist belief systems), patterns, events or sensations associated with the 'object' that is being retained or recalled, or even artefacts such as uniforms, religious symbols, etc., (man-made things to which marker function have been delegated), and which therefore have a degree of 'object-ness' themselves. The total scaffolding structure, that is, the total configuration of relationships among all elements in the network is important in determining the outcome: the actual 'objects' thus defined.

---

[1] I am not referring to modern industrial means, where tools create objects (Ingold 1988).

[2] In my view, the sense in which Lane uses 'scaffolding structure' is a direct extension of the way in which I use it here and refers to its social and material instantiations.

That configuration can be represented, in the language of Actor Network Theory (this volume, Chapters 4, 5, 8), as a network of different actors (human or not), provided one acknowledges that the relationships between human actors and non-human ones are not symmetrical. Humans create the link between 'things' and knowledge that transforms these 'things' into 'objects'. The 'things' or the 'objects' themselves do not do that. Hence, whereas an artefactual 'object' that is (even temporarily) separated from the carrier of its interpretive framework reverts to 'thing-ness', the human being that has created the link can retain a memory of it – the 'object' remains present in memory as 'object', not as 'thing'. Similarly, the different significances that individual 'things' (as 'non-objects') may have in the distributed network of agents are latent and activated by linking these 'things' to knowledge and ideas present in the human mind. Indeed, the links between non-human agents in the network are themselves latent and can only be activated in the human mind.

Next, it is important to underline that the 'mind' in this relationship is itself not an entity. Maybe it is better to speak of 'mindsets'. Individuals have many different cognitive maps in their minds, composed of different 'objects' and relations between them, which are activated differentially according to context and in particular according to the need the individual is attempting to fill at any particular moment. An easy way to see that is by thinking in terms of geographical maps. An individual will have a very different mental map of a particular area, depending on whether he or she is fishing, hiking, driving a car, looking for a site to implant an industrial complex, etc. Each of those maps will include different 'objects' – lakes and waterways in the first case, footpaths, flowers and forests in the second, highways and other drivable roads in the third, and economic conditions, logistics, etc. in the fourth. But one also 'thinks of different things' when eating, repairing a watch, making a pot, etc., in effect possibly associating the same 'thing' with different 'objects'.

The relationship between the different actors in the network (including the relationships between 'things' and the 'objects' that are associated with them in the mind) is a dynamic one. In this dynamic network, association and disassociation between different agents and between different 'things' and 'objects', occurs depending on the person, the mindset, the moment, the context and a range of other variables.

Different 'objects' can therefore be associated with the same 'thing' – as in the case of the sheep in chapter 4 – again, dependent on the function that is being fulfilled, on the purpose of the person involved, on the context, etc. Hence, one can regularly observe mental shifts from one 'object' to another while dealing with the same 'thing', including the creation of associations based on more than one object or of completely new 'objects' related to the same 'thing'.

The core of my argument in this chapter is that invention can profitably be conceived as enabled and constrained by the structure and dynamics of the scaffolding structure that maintains the link between 'things' and 'objects', and that in order to understand the process of invention one has to understand both that structure and the dynamics that maintain it. In other words, to gain an understanding of the process of invention in different instances, we need to investigate the complete network involved, as well as the nature of the dynamic relationships between things and objects and between human and non-human agents.

Of course, the configuration of the network is different in each case and needs to be mapped and studied in terms of network theory, but we can make some assumptions about the general nature of the dynamic relationships involved.

Hypothesis 1: Our ideas ('objects') are under–determined by our observations (and over–determined by our preconceptions).

Our first assumption is that our ideas ('objects') are generally under-determined by our observations, and therefore over–determined by our preconceptions (other 'objects' generated earlier). Atlan (1992) presents an interesting argument to this effect: take five traffic lights, each with three known states (red, orange, green). There are then 243 (3^5) different combinations of the three different colors possible. But there are 25 connections between the traffic lights. That means that there are 3^{25} (about a thousand billion) possible structures (interpretations, theories) that could link the different states of the five traffic lights. Although sequential observations provide more information than random ones, deducing the single correct structure or theory from observations is therefore well–nigh impossible during a lifetime.

That in turn suggests that (1) in observing the same 'thing', different people create somewhat different 'objects', (2) there is a degree of flexibility in the way 'things' are linked with ideas, (3) the context in which the link is made is to some extent determinant for the nature of the resultant 'object' and (4) the ideas to which one links a 'thing' to some extent over-determine the interpretation of that 'thing' as an 'object' in the mind.

If that is so, in many cases our capacity to quickly recognise structure in complex situations must depend on prior knowledge, ways of linking observations and ideas that have been acquired at an earlier time, and proven helpful in such situations. Hence my statement that if ideas are under-determined by observations, we must assume that they are over–determined by earlier ideas.

This hypothesis therefore also links our argument directly to Sutton's in Chapter 3, about the importance of the historical trajectory, and the work of Appadurai and Kopytoff that it is based on. In particular, the historical trajectory of ideas and 'objects' is important, insofar as 'things' are, and remain, poly-interpretable and assume different roles as they are associated with different 'objects'.

Hypothesis 2: The relation between 'things' and 'objects' is determined by the direction of observation

The reduction in dimensions from 'thing' to 'object' is achieved by searching for apparent symmetries (similarities, regularities) between different observations. Our cognitive apparatus allows us to 'fix' instantly certain symmetries in space that disappear the next moment. Repetition of the process also permits us to find and retain temporal symmetries in complex processes with non-harmonic rhythms.

Two different steps can be distinguished in the cognitive process: (a) searching for a perspective that includes a sufficient number of symmetries (defining the problem) and (b) defining the dimensions (finding the solution). Of these, the former takes time and is non-linear because it deals with change, with the uncertainties and risks of the unknown.[3] The latter is instantaneous and linear, fixing as it does a dynamic process into a static perception at the moment at which the future becomes the past. That moment is experienced as the intuitive 'jump' or 'click' of discovery or invention. As soon as a dimension has been defined, it can serve as a (new) "point of view" from which to make observations, providing a new perspective and thus prompting further searches for symmetries. The process is as endless as it is continuous.

Because symmetries need to be distinguished from asymmetries in order to be cognised, the former are continuously compared with the latter and vice versa. Such comparisons can either take asymmetry as subject and symmetry as referent or vice versa. Experimental research on heuristics done by Kahnemann, Tversky and others indicates that the outcome of the comparison is related to the 'direction' of the comparison (e.g., Tversky & Gati, 1978). Comparing a subject with a referent stresses similarities, while comparing a referent with a subject emphasises dissimilarities.

If we apply that idea, which is well-anchored in observational data, to the relationship between 'things' and 'objects', it implies that when 'things' are compared (in the absence of an 'object' to which they are related), that comparison emphasises similarities between the 'things' and identifies the dimensions of similarity. Those dimensions are then stored as an 'object'. Later, when new 'things' are observed and evaluated against that 'object' in order to determine whether they instantiate it, however, this comparison (in the opposite direction) emphasises differences between the 'object' and the 'thing' concerned. As a result, we can identify two phases in the relationship between things and objects: a first one in which the 'object' is as yet 'open-ended', in the sense that it is known which observations might relate to it, but not yet which observations eventually do not, and a second phase, in which the object is 'closed' in the sense that it is known which observations do relate to it, and which do not (Selby & El Guindi, 1976).

If we now relate this to the difference in perspective introduced in the introduction, it is easy to see that cognition of an observed difference between two states of a system at times t_o and t_1 depends on whether the referent is t_1 (the *result* of a presumed process, i.e., when the perception is 'a posteriori') or t_o (the *starting point* of the presumed process as in the case of an 'a priori' perception).

[3] The non-linear nature of the search for a perspective is explained by the fact that in order to cognise change, it is necessary to perceive change in the rate of change.

In my case, the a-posteriori comparison between two states of the system stressed similarity (continuity), produced a scheme of thought or construct ('object') and was formulated in terms of certainties and linear causalities. In Jan Kalsbeek's case, the a-priori comparison was interpreted in terms of a personal experience of time, tended to stress dissimilarity (change) and was cognised in much fuzzier terms, based on concepts such as possibility, risk and uncertainty. Hence our conclusion that *(re-)creating the past from the vantage point of the present is a paradox*. If archaeologists, therefore, wish to reconstruct past decision-making they will have to re-create innovation rather than study it.

Hypothesis 3: The relationship between the world of ideas and the material world is also asymmetric

Another aspect of the asymmetry between the human and the non-human actors in the network manifests itself in the interaction between the world of ideas and the real world. That process is, of course, a two-way street, linking 'things' to 'objects' and 'objects' to 'things'. The former involves a process of *selection* of attributes from among those observed in the real world – the 'object' always has fewer than the (infinite) number of attributes of the 'thing'. In the other direction, *intervention* of humans in the material realm is based on these (simplified, symbolic) 'objects' (ideas), but in connecting them with the material world of 'things', un-cognised attributes of the real world are being involved. This concerns all and any attributes that are connected to the part of the system that is being impacted by human beings. Not only must we therefore conclude that the vast majority of attributes of the outside world remains unknown to us, but also that the instantiation of any human invention in the material realm in due time creates more unknowns. In the domain of sustainability science, these are well-known and called 'unintended consequences'.

Invention in Pottery Making

How does this work out in the 'real world? ' I will try to illustrate that with reference to a network model of the kinds of relations that impact on the capability of a potter to create a set of objects by mediating effectively between the functions of the pottery to be made, the tools and materials at his or her disposal and his or her own know-how, thus relating this part to Malafouris' Chapter 2 of this volume.

The core of this model (Figs 12.2–12.8[4]) is here presented in seven graphs, which together constitute a (rough and incomplete) model of the network of different 'objects' of the pottery making system that the potter has to keep in mind if the pottery is in effect to be realised. These 'objects', in turn, are associated in multiple ways with different phenomena ('things') in the real world. The exact nature of the relationships between these 'objects', or between

[4] Figures 2.8 and table 2 have been adapted from van der Leeuw 1980.

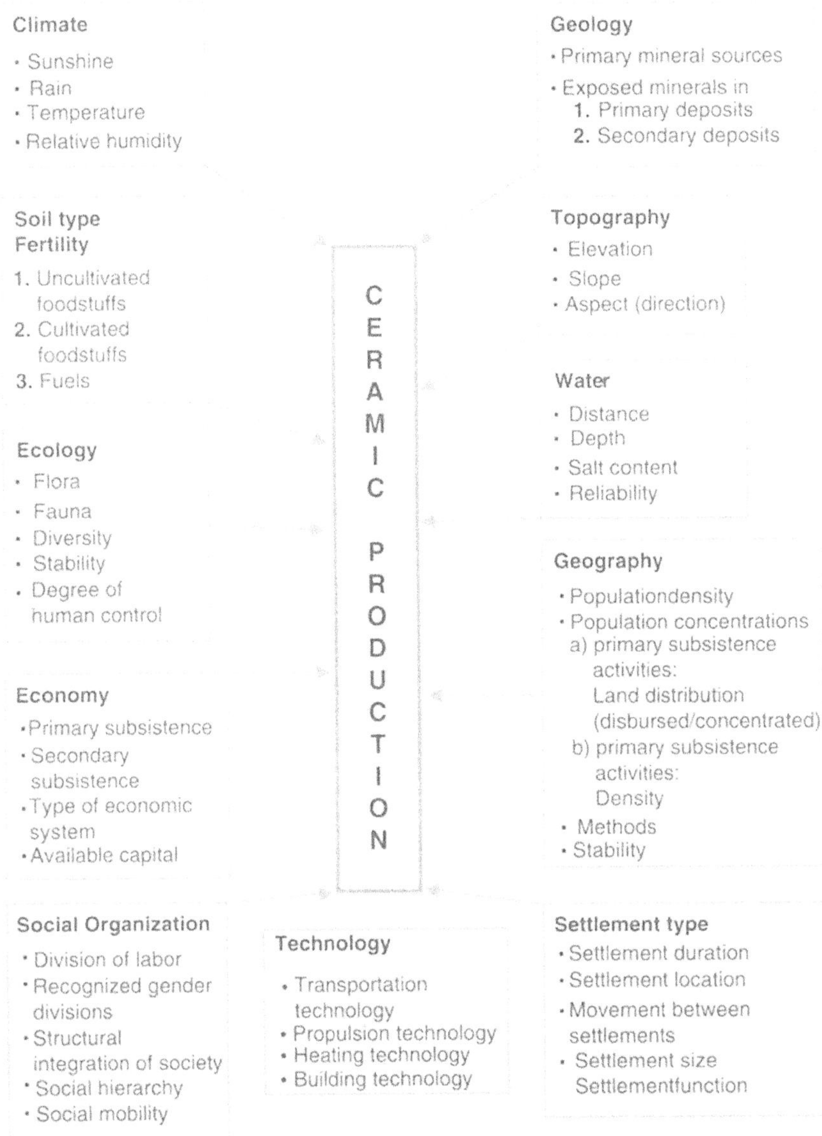

Fig. 12.2 When making pottery, the potter needs to take a wide range of aspects of the context of production into account, including environmental conditions, geography, aspects of the society, etc.

the 'things' and the 'objects', has not been elaborated in these figures, as it differs from case to case. To complicate matters, each of these 'objects' can be in different states: the clay can be dry or wet, too plastic or not plastic enough, the firing can be insufficiently hot, or too hot, etc. Each of the seven figures (Figs. 12.2–12.8) highlights a different perspective on the network that governs

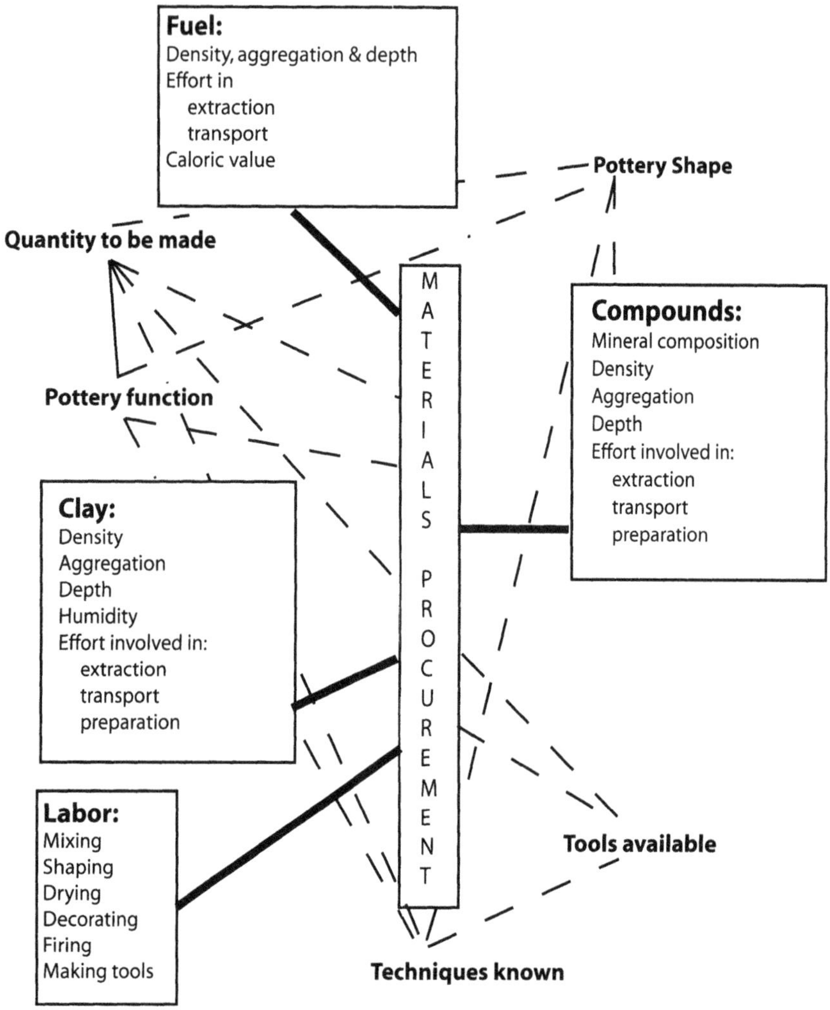

Fig. 12.3 Network representation emphasising certain aspects ('objects') impacting on raw material procurement (in boxes in the figure) that the potter has to keep in mind in order to obtain suitable raw materials for his work at a convenient cost

pottery making: the first figure, Fig. 12.2 presents some of the elements of the context of pottery making which every potter has to take into account. Figures 12.3–12.4 deal with some of the variables important for raw materials procurement and preparation, Figs. 12.5 and 12.6 with some of the elements factored in by the potter when conceptualising and shaping the pottery, Fig. 12.7 with some of the factors determinant for marketing and pricing, and Fig. 12.8 with aspects of the organisation of the workshop.

Of course, each figure presents a partial perspective on the total network of 'things' and 'objects' and when making pots, the potter has to keep the content of all these figures (and a mass of others) in mind. The 'objects' are elaborated in

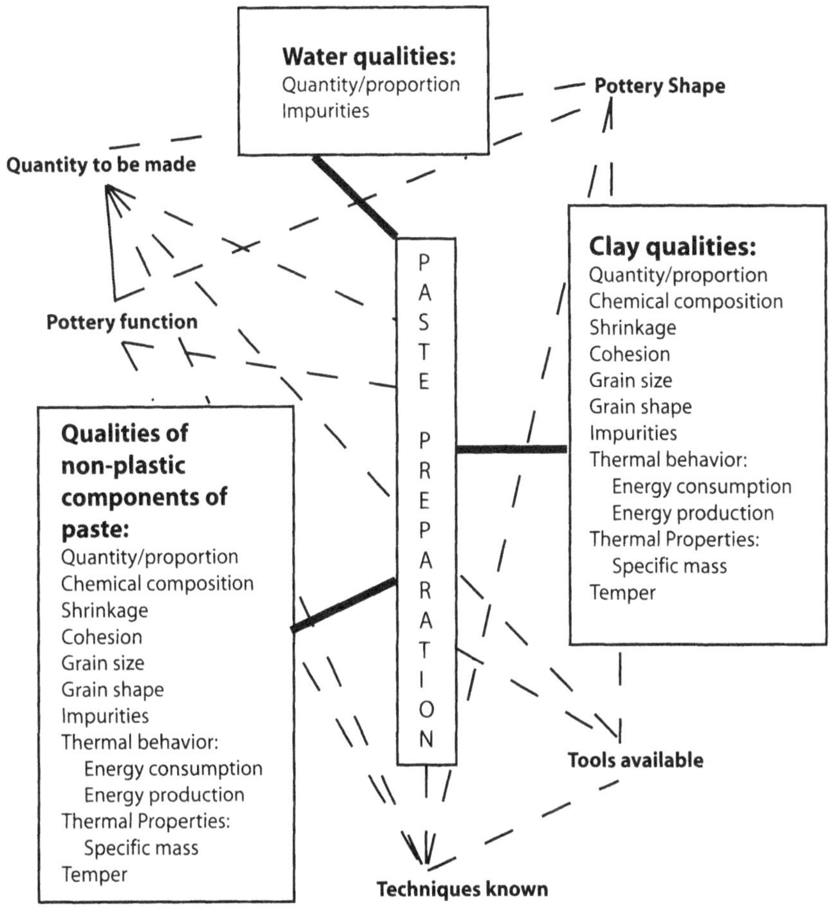

Fig. 12.4 Network representation emphasising the 'objects' (in boxes in the figure) impacting on the potter's transformation of raw materials into a suitable paste to use in pottery making

the kind of asymmetrical interactions of the mind with the real world that we have discussed above, and all of them (as well as others) figure in the network that governs pottery making together with the 'things' with which they have (often multiple) relations of perception and instantiation.

Complex though the network is, interestingly *the number and nature of the relationships in it is limited by the nature of the real world*, which does validate certain relations, but does not tolerate others: dry clay cannot be molded or otherwise shaped; certain shapes cannot be realised due to the laws of physics; each clay requires a minimum amount of heat to be fired, but cannot be fired beyond a certain temperature, etc.

Other limitations derive from the economic context in which the production occurs. This is represented in Table 12.2, which describes pottery making in terms of a series of system states constrained by the size of the production, and thus by the market conditions in which the work is done.

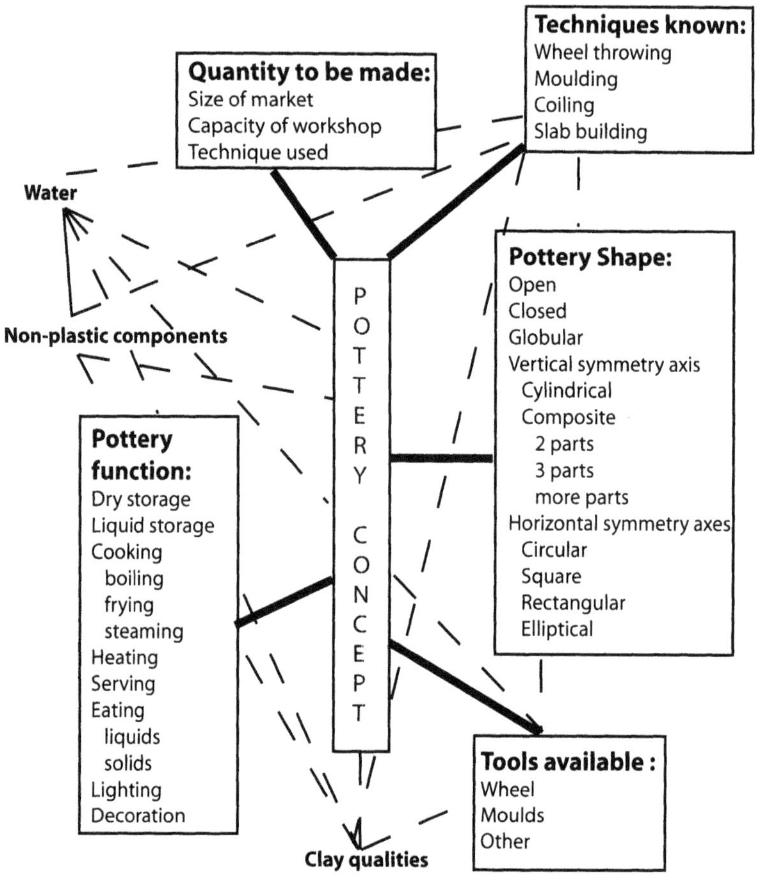

Fig. 12.5 Network representation emphasising some of the technical 'objects' the potter keeps in mind while conceptualising the pots to be shaped (in boxes in the figure)

Even then, these limits to the total size and structure of the network, however, far exceed the actual part of the network that is activated in routine production within any one manufacturing tradition. *The third limitation is inherent in the history of the tradition (the network).* Once certain solutions have been found (i.e., once certain 'things' have been transformed into 'objects', and once it has subsequently been proven that these 'objects' can successfully be instantiated into more 'things'), the sheer complexity of the network and of the instantiation procedures favours the continued use of these 'solutions' over the invention of new ones as long as the context is the same. As we will see below, for example, once it became a habit on Negros to begin the vessels by shaping the rim, this was maintained throughout a range of ulterior modifications of the manufacturing procedure, such as the introduction of more and more sophisticated rotation devices. But ultimately, that same choice to begin with the rim was not compatible with the use of the modern, western, potter's wheel, and that was therefore difficult to introduce.

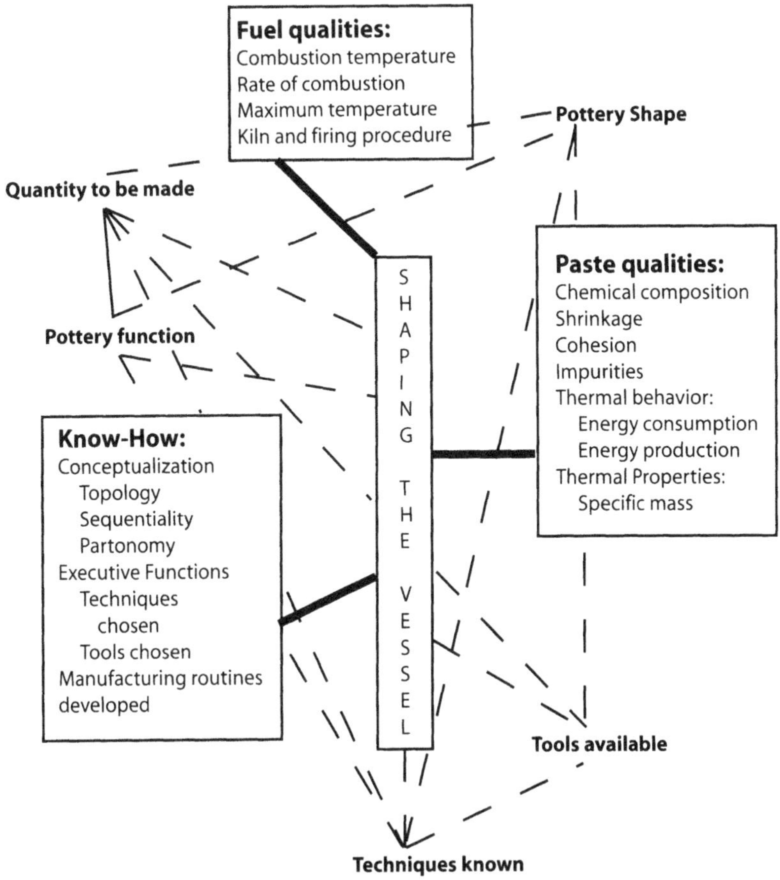

Fig. 12.6 Network representation emphasising some of the technical 'objects' the potter has to keep in mind while shaping the vessels (in boxes in the figure)

Suppose the potter has in mind to make a jar to pour water from. That imposes certain minimal constraints both on the shape of the vessel (a high centre of gravity, a handle, a spout) and on the characteristics of the vessel wall (it has to be waterproof). The tools he or she uses impose other constraints: on a potter's wheel, *ceteris paribus*, there is more potential variation in shape than when the potter is using a mould, for example. Other constraints derive from the nature of the kiln and fuel: if the kiln can be fired above 1050 C, the potter can usually ensure impermeability of the vessel wall by vitrification, but if that is not the case, the vessel either has to be glazed or impregnated with another substance to make it impermeable. Yet other constraints derive from the availability of raw materials: if the clay is fine and plastic, and the potter is working on a wheel, the clay may be used as is, but if it is found with larger non-plastic admixtures, it will have to be purified. On the other hand, if the potter is working in a mould, the 'rougher' paste may be used, and a fine clay will have to be mixed with some form of temper, etc.

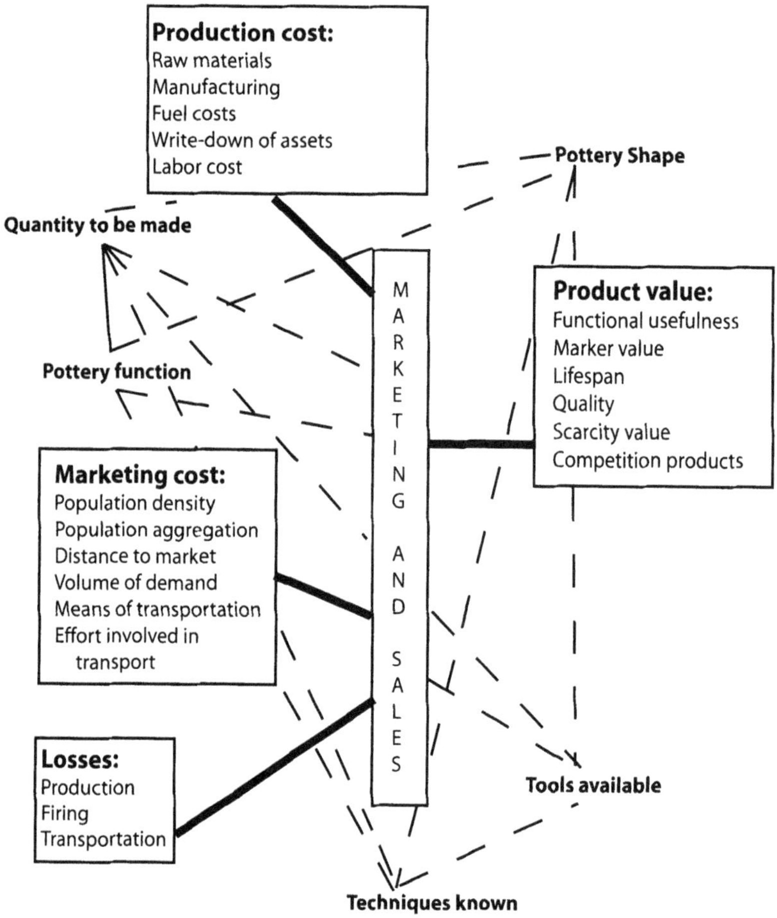

Fig. 12.7 Network representation of some of the aspects ('objects') of the market and the marketing organisation that impact on the marketing of the vessels (in boxes in the figure)

These are but some examples to illustrate the nature and complexity of the process by which the potter articulates all the different elements to be taken into account to successfully produce the desired vessel. In reality, many more such elements and considerations come into play, some of which are referred to in the illustrations. But the main points to make here are (1) that this process is subject to the three asymmetries that we have outlined earlier, (2) there are always multiple 'solutions' in the general model to realise one's aim and (3) the historical trajectory of a particular 'tradition' generally has limited the choice of potential 'solutions' of which the potter is aware by circumscription of the set of 'objects' which constitutes the 'know-how' of the artisan.

At the level of the individual relations between people, 'things' and 'objects', the dynamics are subject to the three working hypotheses outlined above: the asymmetries between cognition and action, the asymmetry between 'things' and

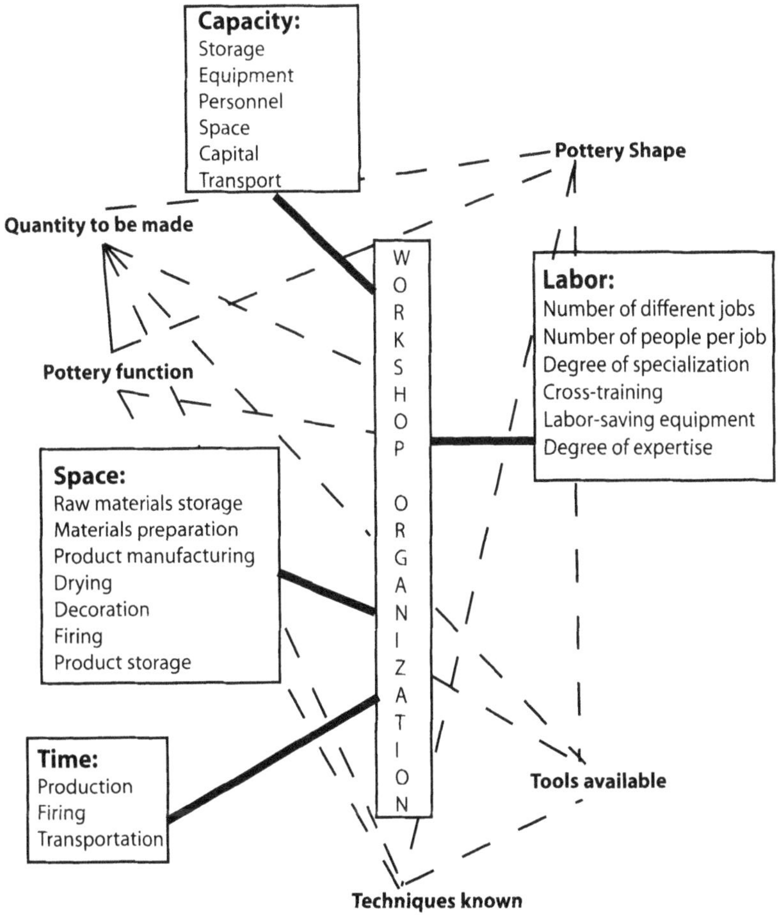

Fig. 12.8 Network representation of aspects of the organisation of the work and the workshop ('objects' in boxes in the figure) that impact on a potter's decisions in planning and implementing production

'objects' and the under-determination of our ideas by observations (and the concomitant over-determination by understanding or know-how acquired earlier).

As a result of all these asymmetries, the network that connects the nodes is to some extent 'flexible'. There is flexibility in the connections between 'things' and 'objects' in the sense that certain things do not always have completely the same associations in the mind – because those associations are in part dependent on the context, as well as on the direction in which the different nodes are connected. But it is important to emphasise that this does not mean that the network is 'unstructured'. Rather, the asymmetries of the cognitive relations create a kind of 'structure of biases' which over time impact on the trajectory of invention in a society.

Table 12.2 Relationship between the economics and the technology of pottery making. The table presents a number of 'system states' of the overall pottery making system in order to show how production (and sales) volume relate to the organisation of pottery making, to the techniques used, and to the raw materials selected for use

	variables	household production	household industry	individual industry	workshop industry	village industry	large-scale industry
Economy	Time involved	Occasional	Part-time	Full-time	Full-time	Part-time/Full-time	full-time
	Number involved	One	Several	One	Several	several	Many
	Organization	None	None	None	(Guild)	Certain	Certain
	Locality	Sedentary or itinerant	Sedentary or itinerant	Itinerant	Sedentary	Sedentary	Sedentary
	Hired hands	None	None	None	Some	Some	Labour force
	Market	Own use	Group use	Regional	Village/town	Region (wide)	Regional and export
	Raw materials						
	Clay	Local	Local	Local	Neighbourbood	Neighbourhood	Neighbourbood/distant
	Temper	Local	Local	Local	Neighbourhood	Neighbourhood	Neighbourhood/distant
	Water	Local	Local	Local	Local	Local	Local
	Fuel	Local	Local	Local	Neighbourhood	Neighbourhood	Neighbourhood/distant
	Investments	None	None	Few	Some	Some	Capital
	Seasonality	Production as needed	Season w/o other work	All year except winter	All year/good weather	All year/good weather	All year
	Labour division	None	None	None	Some-considerable	Some-considerable	Detailed
	Time involved per Pot	High	High	Medium	Medium-low	Medium-low	Low
	Status	Amateur	semispecailist	Specialist	specialist	Specialist	Specialist(few techniques)

Table 12.2 (continued)

	variables	household production	household industry	individual industry	workshop industry	village industry	large-scale industry
Technology	Manuf. techniques tools	Hand/small tools	Hand/small tools	Hand/small tools	Mould/wheel	Mould/wheel	Wheel/cast/press
	Sed. Basin	None	None	None	When needed	When needed	Needed
	Wheel	None	None; rotary support	Turntable	Various kinds	Various kinds	Kickwheel or similar
	Drying shed	None	None	None	Needed	Needed	Needed
	Kiln	Open firing	Open firing/impermanent	Impermanent	(semi-) permanent	(semi-) permanent	Permanent
	Raw materials						
	Clay	Wide range	Wide range	Wide range	Narrower range	Narrower range	Narrow range
	Temper	Wide range	Wide range	Wide range	Narrower range	Narrower range	Narrow range
	Water	Any	Any	Any	Any	Any	Any
	Fuel	Wide range	Wide range	Wide range	Narrower range	Narrower range	Narrow range
	Range of pottery	Narrow	Narrow	Wide	Narrow or wide	Narrow or wide	Narrow or wide
	Range of functions per pot	wide	wide	wide	narrower	narrower	Narrower
Examples					Bergen-op-Zoom	Tzintzuntzan	Wedgwood
		Kabyles, N Africa	Cameroon, Tanzania	Tibet	Farnham	Temascalcingo	Delft
					Haarlem	Djerba	

There are two sets of circumstances under which the artisan has to adapt by making changes in the mediation process that underpins production. The first of these is when unintended consequences of certain choices made in the production process lead to problems or incompatibilities within the specific model underlying manufacture so that that model cannot satisfactorily be applied and the other is when any of the parameters governing a particular way of making a category of artefacts changes (new economic conditions, new materials, new shape (or other) requirements, etc.).

When either set of circumstances occurs, the artisan has to re-evaluate all the elements of the specific conceptual model on which production is based and the relations between them. That process takes place from the 'a priori' perspective that is the artisan-creator's and is subject to the above-mentioned biases inherent in the cognitive articulation of the world of ideas with the material world. Depending on the situation, and in particular on the nature and extent of the problems that emerge, this will lead to either incremental improvement or to drastic change such as the invention of completely new techniques. But in either case, the 'new' emerges from the pre-existing network of 'agents' by looking at that network 'anew', devising new conceptualisations, new ways to mediate between the different materials, tools, ideas and know-how that enable the creation of artefacts. And over the longer term, the 'structure of biases' mentioned above therefore inevitably has an impact on the evolution of invention in the system, creating a 'tradition'.

Innovation

The core of my argument in this chapter is that invention and innovation play out in a dynamic network relating agents, things, 'objects' and contexts. In such a network, invention is a local process, involving few nodes and few edges. Only those agents that are immediately connected to the inventor(s) are involved and the process is therefore one that plays out in a relatively limited number of dimensions comprising the immediate articulation of the know-how of the inventor(s) with the other agents in their material world.

Innovation however, the process by which an invention spreads and is taken up – and may even generate an 'innovation cascade' – plays out in a much larger network of agents. The network dynamics involved are of several different kinds. First of all, there is variability in the extent of, and the parts of, the scaffolding structure that are activated at any one time, so that new associations can be made, or old ones 'lost'. This variability is related to the topology of the network itself – that is, to the nature and configuration of the nodes in it and the edges between them. But it is also related to the context(s) of activation and to the accumulated understanding that the human actors in it have gained in the past as part of the activation of (partially) different networks. But that accumulated understanding is not activated in its entirety: it is activated in the form of associations that the human agents make between a current activation and anterior ones.

This may be illustrated by contrasting two ceramic traditions, pottery making in a mould in Michoacan (Mexico; van der Leeuw, Papousek & Coudart, 1992; van der Leeuw & Papousek, 1992), which I have studied in the early 1990s and pottery making on different forms of rotating devices, such as I have studied on Negros (Philippines) in the early 1980s (van der Leeuw, 1984a, b). As limitations of space do not permit me here to work these examples out in great detail, or even to map the networks completely, all I can do here is to present the differences between the two systems in tabular form and add some explanation. A more elaborate presentation is in preparation.

The two examples essentially represent workshop industries in which many potters in the same location compete in making a range of pots in considerable numbers for sale in a market economy.[5]

What immediately strikes one in Table 12.3 is that the *social context* of pottery making in the two locations is very different, and does indeed seem much more conducive to invention and innovation in the Philippines than in Mexico because it favours information exchange in the former but works against it in the latter. But although that may explain *why* innovation was more manifest in the former than in the latter society, it does not explain *how* it occurred in either society.

Following on from the social context, the *organisation of production* is more flexible on Negros. In both cases, men and women make pots, although on Negros men and women can both own property and are economically more independent and equal than in Michoacan, where the society is strongly dominated by men. In Michoacan, the family unit is essentially the basis of the organisation of pottery making, whereas in the Philippines, it is a commercial unit that may include (paid) workers that do not belong to the family. This does not make any important difference in the organisation of the work itself but it does enable the workshops on Negros to adapt their organisation more quickly to new needs.

Let us then look at the *toolkits*. The tools used in the original traditions are the paddle-and-anvil and the mold respectively. But in the Negros tradition, the use of the mold as well as of a rotating device has been introduced, whereas in Michoacan, the mold has remained the principal tool, with very few exceptions. On Negros, it is common to mount a half horizontal mould (an upside-down

[5] Although this is the case for the vast majority of workshops on Negros, the full range of organisational forms is represented, from the single potter who only makes pottery every once in a while via the single individual who makes pottery full time, all the way to the workshop industry and the full industrial factory. Indeed, it is a characteristic of the Negros case that the potters have been able to adapt their toolkit, know-how and organisation in a relatively short period to mass production by inventing new technical and organisational solutions to the challenges posed by the original pottery making tradition. The Michoacan potters have not done so.

Table 12.3 Some of the principal differences between pottery making traditions on Negros and in Michoacan

Negros	Michoacan
Natural environment	
Tropical	Mediterranean
Social context	
Very interactive society	Closed, fragmented society
Pottery making in open air, visible to all	Walls around workshops or pots made inside
Individuals can undertake change	Change requires collaboration of many people
Everyone has all know-how to effect change	Only moldero has know-how to effect change
No visible social means to limit change	Numerous social means to limit change
Makers' conceptualisation	
Makers' topology does not distinguish inside and outside of vessels	Makers' topology fundamentally distinguishes inside and outside of vessels
Makers' partonomy distinguishes shapes made as wholes or shapes made in horizontal segments	Makers' partonomy distinguishes horizontally molded open shapes or closed shapes molded in vertical segments
Makers' sequence is lip to base	Makers' sequence is in vertical segments
Neither shapes nor decoration prescribed	Shapes prescribed, decoration partially proscribed
Tools and executive functions	
Basic tradition is hammer and anvil shaping	Basic tradition is mold-shaping
Molds used as aids for certain shapes	Molds essential for all shapes
Rotating devices used as aids in many workshops	Rotating devices used only in two a-typical workshops, for unusual reasons
Firing principally in open air, exceptionally in kiln	Firing in controlled environment (kilns)
Organisation of production	
Wide range, from 5-6 to thousands of pots per month per workshop	Thousands of pots per month per workshop
Varies with size of production; full range from individuals via family workshops to industrial workshops and cooperatives	(Extended) Family – based workshops
Division of labour in larger workshops	Only informal, temporary division of labour within the family
Raw materials	
Clays and temper from different locations, mixed at the workshop	Clay deposits generally contain enough non-plastic materials to avoid using separate temper

pot) on a stick and to allow it to rotate in an upright bamboo tube, while in Michoacan, very occasionally, a similar device is being used to support the heavy horizontal mold on which one shapes plates, bowls and other 'open' objects. Although in both places there are people who have attempted to make the potters aware of the European kickwheel, their attempts have not been successful in either region. Finally, kilns are known in both places, but while they are commonly used in Michoacan (where fuel is scarce), they are only rarely used in Negros (where fuel is plentiful). In view of all that, contrary to habitual interpretations in archaeology, the difference in toolkit itself can therefore not be held responsible for the different ways in which innovation has manifested itself in the two traditions.

As for the *raw materials*, these are pretty much the same in both regions – whether the clay is dug up mixed with sandy temper or whether that temper is added. The main difference is that in Michoacan the clay is dug dry almost all year around, then ground and reconstituted into a plastic mass by adding water, whereas on Negros, a large part of the year, the clay is dug wet, so that grinding is avoided. In the industrial workshops on the island, sand is then added.

We conclude that none of the above categories of attributes of the pottery making system seem individually to explain the difference in direction of inventiveness and innovation that we observe when comparing the two regions. However, brought together into an interactive triangle between people, ideas and material 'things', they may provide part of the story. The clue to all that is found in the interaction between ideas ('objects' in the terms of this chapter) and the 'things' in the material world, in the context of the social-organisational systems in the two areas.

Let us therefore turn to the makers' *conceptualisations*. In the case of Negros, pots are conceived in horizontally joined parts. In most cases, the rim is shaped first and the body next. Whether the shaping of the body is by hammer-and-anvil out of a single lump of paste or by shaping the base and the shoulder separately and joining them together, the conception of the pot is based on a horizontal partonomy and a sequence that begins with the rim. Such a conception has facilitated the introduction of a vertical rotary device. In Michoacan, on the other hand, the more complex pots are conceived of in terms of a vertical partonomy – the pots are shaped in molds that consist of two or more vertical sections. Such a vertical partonomy evidently does not facilitate the use of a rotary device turning around a vertical axle.

As in every tradition, ultimately ideas, tools and know-how become closely aligned and intertwined into a system in which changes in one domain immediately require changes in many others. Thus anchored, the two traditions developed in very different ways – on Negros, a rapid increase in production was facilitated by better and better turning devices, but the range of shapes that could be made on them remained limited. The fact that pots are conceived from the rim down blocked the introduction of a western-style potter's wheel. In Michoacan, increasing production could not be achieved by increasing productivity and the vertical conception of the vessels completely excluded the introduction of a western-style wheel. Mold-making is in effect relatively efficient, but increased production required more people and more moulds. On the other hand, the vertical partonomy enabled the making of a very wide range of shapes, including annular pots, pots in the shape of birds, and asymmetric pots, which would never have been 'invented' on Negros.

The need for mold-makers and potters to agree on the creation of new shapes, and the fierce inter-village competition, coupled with the 'closed' social structure, hampered innovation in shaping techniques in Michoacan. Creativity was much more evident in decoration, where many traditional motifs were in the process of being exchanged against much more innovative ones, because in this domain there were no physical constraints against change, and the social rules were less stringently formulated.

Conclusion

In this very preliminary study, I have argued that we urgently need to understand the process of invention and innovation so as to avoid always re-acting to innovations, instead of harnessing innovation to our needs. Thus far, such understanding has been hampered by what I have here called an 'a posteriori', scientific, approach to the topic. Instead, we need to develop an 'a priori' one, which looks at the logic of creation in terms of choices made and choices not made. In developing such an approach, the 'distributed' network approach to agency developed in this book is very helpful, provided one does not forget that humans are the medium through which action occurs and that, therefore, the inherent asymmetries of human interaction with the material world play an important role. These asymmetries are over time responsible for a 'structure of biases' in each and every technological tradition which impact on choices made and ultimately, on which kinds of inventions and innovations are affordable in a tradition.

To gain this kind of understanding of invention and innovation, it is important to consider – from a network perspective – all the factors that may play a role in the decision-making process, including social and environmental ones.

References

Appadurai, A., 1986, Introduction: Commodities and the Politics of Value. In *The Social Life of Things: Commodities in Cultural Perspective*, edited by A. Appadurai, pp. 3–63. Cambridge University Press, Cambridge.

Atlan, H., 1992, Self-organizing Networks: Weak, Strong and Intentional. The Role of Their Underdetermination. *La Nuova Critica*, N.S., 19–20(1/2): 51–70.

Borges, J.-L., 1981, Tlön, Uqbar, Orbis Tertius (Translation A. Bonner), in *Ficciones*, edited by A. Kerrigan, 17–37. Grove Press, New York.

Brown, B., 2003, *A Sense of Things: The Object Matter of American Literature*. University of Chicago Press, Chicago.

Callon, M., 1986, Some Elements for a Sociology of Translation: Domestication of the Scallops and the Fishermen of St-Brieuc Bay. In *Power, Action and Belief: A New Sociology of Knowledge?*, edited by J. Law, pp. 196–223. Sociological Review Monograph, Routledge and Kegan Paul, London.

Girard, R., 1990, Innovation and Repetition. *SubStance* 62/63: 7–20.

Gosden, C., 2004, Aesthetics, Intelligence and Emotions: Implications for Archaeology. In *Rethinking Materiality: The Engagement of Mind with the Material World*, edited by E. DeMarrais, C. Gosden and C. Renfrew, pp. 33–43. McDonald Institute for Archaeological Research, Cambridge.
Harrison, S., Pile, S., and Thrift, N., editors, 2004, *Patterned Ground: Entanglements of Nature and Culture*. Reaktion Books, London.
Ingold, T., 1988, Tools, Minds and Machines: An Excursion in the Philosophy of Technology. *Techniques et Culture* 12: 151–76.
Kopytoff, I., 1986, The Cultural Biography of Things: Commoditization as Process. In *The Social Life of Things: Commodities in Cultural Perspective*, edited by A. Appadurai, pp. 64–91. Cambridge University Press, Cambridge.
Latour, B., 1993, *We Have Never Been Modern*. Harvard University Press, Cambridge, MA.
Malafouris, L., 2004, The Cognitive Basis of Material Engagement: Where Brain, Body and Culture Conflate. In *Rethinking Materiality: The Engagement of Mind with the Material World*, edited by E. DeMarrais, C. Gosden and C. Renfrew, pp. 53–62. McDonald Institute for Archaeological Research, Cambridge.
McGlade, J., and van der Leeuw, S.E., 1997, Archaeology and Nonlinear Dynamics: New Approaches to Long-term Change. In *Archaeology: Time, Process and Structural Transformations*, edited by S.E. van der Leeuw and J. McGlade, pp. 1–31. Routledge, London.
Prigogine, I., 1980, *From Being to Becoming*. W.H. Freeman, San Francisco, CA.
Searle, J.R., 1983, *Intentionality: An Essay in the Philosophy of Mind*. Cambridge University Press, Cambridge.
Selby, H., and El Guindi, F., 1976, Dialectics in Zapotec Thinking. In *Meaning in Anthropology*, edited by K.H. Basso and H.A. Selby, pp. 181–96. School of American Research Advanced Seminar Series, University of New Mexico Press, Albuquerque:
Tversky, A., and Gati, I., 1978, Studies of Similarity. In *Cognition and Categorisation*, edited by E. Rosch and B.B. Lloyd, pp.79–98. Erlbaum, Hillsdale, NJ.
Van der Leeuw, S.E., 1976, *Studies in the Technology of Ancient Pottery*, 2 vols., Unpublished Thesis, University of Amsterdam.
Van der Leeuw, S.E., 1980, Keramikproduktion und Keramikhandel. Methodologische Probleme Ihrer Erforschung. *Luebecker Schriften zur Archaeologie und Kulturgeschichte* 4:185–206.
Van der Leeuw, S.E., 1984a, Manufacture, Trade and Use of Pottery on Negros, Philippines. In *Earthenware in Asia and Africa*, edited by J. Picton, pp. 326–364. Colloquies on Art and Archaeology in Asia, 12, Percival David Foundation of Chinese Art, London.
Van der Leeuw, S.E., 1984b, Dust to Dust: A Transformational View of the Ceramic Cycle. In *The Many Dimensions of Pottery: Ceramics in Archaeology and Anthropology*, edited by S.E. van der Leeuw and A.C. Pritchard, pp. 707–773. Amsterdam.
Van der Leeuw, S.E., 2007, Information Processing and Its Role in the Rise of the European World System. In *Sustainability or Collapse?*, edited by R. Costanza, L.J. Graumlich, and W. Steffen, pp. 213–241. Dahlem Workshop Reports, MIT Press, Cambridge, MA.
Van der Leeuw, S.E., Papousek, D.A., and Coudart, A.G., 1992, Technological Traditions and Unquestioned Assumptions. *Techniques et Culture* 17–18: 145–173.
van der Leeuw, S.E., and Papousek, D.A., 1992, Tradition and Innovation. In *Ethnoarchéologie: Justification, Problèmes, Limites*, edited by F. Audouze, A. Gallay and V. Roux, pp. 135–158. A.P.D.C.A., Antibes.

Index

T

Printed in the United States of America